ADVANCED BIOFUELS

Using Catalytic Routes
for the Conversion of
Biomass Platform Molecules

ADVANCED BIOFUELS

Using Catalytic Routes
for the Conversion of
Biomass Platform Molecules

Edited by

Juan Carlos Serrano-Ruiz, PhD

APPLE
ACADEMIC
PRESS

Apple Academic Press Inc.	Apple Academic Press Inc.
3333 Mistwell Crescent	9 Spinnaker Way
Oakville, ON L6L 0A2	Waretown, NJ 08758
Canada	USA

©2016 by Apple Academic Press, Inc.

First issued in paperback 2021

Exclusive worldwide distribution by CRC Press, a member of Taylor & Francis Group
No claim to original U.S. Government works

ISBN 13: 978-1-77463-557-5 (pbk)
ISBN 13: 978-1-77188-132-6 (hbk)

Library and Archives Canada Cataloguing in Publication

Advanced biofuels: using catalytic routes for the conversion of biomass platform molecules / edited by Juan Carlos-Serrano Ruiz, PhD.

Includes bibliographical references and index.
ISBN 978-1-77188-132-6 (bound)
1. Catalysts. 2. Catalysis. 3. Biomass energy. I. Serrano-Ruiz, Juan Carlos, editor

TP159.C3A38 2015	660'.2995	C2015-900407-1

Library of Congress Cataloging-in-Publication Data

Advanced biofuels: using catalytic routes for the conversion of biomass platform molecules/editor, Juan Carlos Serrano Ruiz, PhD.

pages cm
Includes bibliographical references and index.
ISBN 978-1-77188-132-6 (alk. paper)
1. Biomass energy. 2. Catalysis. 3. Green chemistry. I. Serrano-Ruiz, Juan Carlos.

TP339.A367 2015	662'.88--dc23	2015000864

Apple Academic Press also publishes its books in a variety of electronic formats. Some content that appears in print may not be available in electronic format. For information about Apple Academic Press products, visit our website at **www.appleacademicpress.com** and the CRC Press website at **www.crcpress.com**

About the Editor

JUAN CARLOS SERRANO-RUIZ, PhD

Dr. Juan Carlos Serrano-Ruiz is licensed in Chemical Sciences by the University of Granada. He received his PhD in Chemistry and Material Science from the University of Alicante, Spain. He has visited many laboratories all around the world in his research on biofuel. He was a MEC/Fulbright Student at the University of Wisconsin-Madison, where he studied catalytic conversion of biomass. Upon his return to Spain, he accepted work at the Department of Organic Chemistry at the University of Cordoba, where he has continued his work with biofuels. He is the author of more than fifty scientific publications in international journals, including an article in *Science Magazine* on using sugar as a biofuel. He is also the coinventor of a patent taken out by the Wisconsin Alumni Research Foundation for the conversion of cellulose into diesel and gasoline.

Contents

Acknowledgment and How to Cite

The editor and publisher thank each of the authors who contributed to this book. The chapters in this book were previously published in various places in various formats. To cite the work contained in this book and to view the individual permissions, please refer to the citation at the beginning of each chapter. Each chapter was read individually and carefully selected by the editor; the result is a book that provides a nuanced look at the possibilities of a new generation of biofuels. The chapters included are broken into three sections, which describe the following topics:

- Chapter 1 describes some breakthroughs in analytical tools and synthetic approaches toward improved energy efficiency and catalyst stability.
- As a part of a basic overview of this topic, Chapter 2 looks at a range of processes and applications.
- Again as a part of a basic overview on the book's topic, Chapter 3 offers insights into the various analytical methods that are useful for evaluating the efficiency of catalytic reactions in the transformation of biomass into usable fuel.
- The authors of Chapter 4 investigate the aqueous-phase routes used for the reactions that convert sugars into liquid hydrocarbons, focusing on the sustainability of using a small number of reactors with minimum use of hydrogen sources from fossil fuels.
- The authors of Chapter 5 select phenol, water, acetic acid, acetaldehyde hydroxyacetone, D-glucose, and 2-hydroxymethylfuran as typical bio-oil components and mixed them as a synthetic bio-oil, in order to demonstrate some of the competing reaction pathways that occur in bio-oil upgrading by acid-catalyzed alcohol/olefin treatment.
- The authors of Chapter 6 take a novel approach to the problem of high emissions during small-scale biomass production by investigating the practical use of catalytic components in a downdraft wood stove.
- The research in Chapter 7 considers some cross-coupling reactions that use relatively economic nucleophilic partners. The authors propose that this type of catalysis might be useful for extracting some high-value products from bio-oil mixtures, and they identify several new protocols for cross-coupling.

- The authors of Chapter 8 describe an optimized biphasic system that could aid the development of a simple and cost-effective protocol for the conversion of various carbohydrates. Their results offer several advantages over some of the other methodologies, including mild reaction conditions, satisfactory product yields, and a simple isolation process.
- The recyclability experiments in Chapter 9 indicate that sulfonated Periodic Mesoporous Organosilica (PMO) is a reusable and stable option as a catalyst in biofuel productions.
- Chapter 10 enhances our understanding of the use of ultrasonic sound waves to accelerate the transesterification process, which could potentially lead to substantial improvement in both batch and continuous production systems, thus making biomass conversion a more sustainable process.

List of Contributors

Dimitris S. Argyropoulos
Departments of Chemistry and Forest Biomaterials, North Carolina State University, Raleigh, NC 27695-8005, USA and Center of Excellence for Advanced Materials Research (CEAMR), King Abdulaziz University, P.O. Box 80203, Jeddah 21589, Saudi Arabia

Stella Bezergianni
Chemical Processes & Energy Resources Institute (CPERI), Centre for Research & Technology Hellas (CERTH), Thermi-Thessaloniki, Greece

R. Bindig
DBFZ Deutsches Biomasseforschungszentrum gemeinnützige GmbH, Torgauer Straße 116, Leipzig 04347, Germany

S. Butt
DBFZ Deutsches Biomasseforschungszentrum gemeinnützige GmbH, Torgauer Straße 116, Leipzig 04347, Germany

Kefu Chen
State Key Laboratory Pulp and Paper Engineering, South China University of Technology, Guangzhou 510460, China

Matthew L. Clarke
School of Chemistry, University of St Andrews, EaStCHEM, St Andrews, Fife, UK.

Els De Canck
Center for Ordered Materials, Organometallics & Catalysis (COMOC), Department of Inorganic and Physical Chemistry, Ghent University, Krijgslaan 281, Building S3, Ghent B-9000

Inmaculada Dosuna-Rodríguez
Institute of Condensed Matter and Nanosciences (IMCN)—Division "MOlecules, Solids and reactiviTy—MOST", Université catholique de Louvain, Croix du Sud 2/L7.05.17, Louvain-la-Neuve B-1348, Belgium.

James A. Dumesic
Department of Chemical and Biological Engineering, University of Wisconsin–Madison, USA

Eric M. Gaigneaux
Institute of Condensed Matter and Nanosciences (IMCN)—Division "MOlecules, Solids and reactivi-Ty—MOST", Université catholique de Louvain, Croix du Sud 2/L7.05.17, Louvain-la-Neuve B-1348, Belgium.

Wenhua Gao
Departments of Chemistry and Forest Biomaterials, North Carolina State University, Raleigh, NC 27695-8005, USA and State Key Laboratory Pulp and Paper Engineering, South China University of Technology, Guangzhou 510460, China

Gavin J. Harkness
School of Chemistry, University of St Andrews, EaStCHEM, St Andrews, Fife, UK. E-mail: mc28@st-andrews.ac.uk; Fax: +44 (0)1334 463808; Tel: +44 (0)1334 463850

I. Hartmann
DBFZ Deutsches Biomasseforschungszentrum gemeinnützige GmbH, Torgauer Straße 116, Leipzig 04347, German

Bjarne Holmbom
Åbo Akademi University, Process Chemistry Centre, Finland

Stuart M. Leckie
School of Chemistry, University of St Andrews, EaStCHEM, St Andrews, Fife, UK. E-mail: mc28@st-andrews.ac.uk; Fax: +44 (0)1334 463808; Tel: +44 (0)1334 463850

Adam F. Lee
European Bioenergy Research Institute, Aston University, Aston Triangle, Birmingham, B4 7ET, UK

Yiqun Li
Department of Chemistry, Jinan University, Guangzhou 510632, China

Dmitry Murzin
Åbo Akademi University, Process Chemistry Centre, Finland

José Ortiz-Landeros
Departamento de Ingeniería en Metalurgia y Materiales, Escuela Superior de Ingeniería Química e Industrias Extractivas, IPN, UPALM, México DF, Mexico

Heriberto Pfeiffer
Instituto de Investigaciones en Materiales, Universidad Nacional Autónoma de México, Circuito exterior s/n, Ciudad Universitaria, Del. Coyoacán, México DF, Mexico

Charles U. Pittman, Jr.
Department of Chemistry, Mississippi State University, Mississippi State, MS 39762, USA

Margarita J. Ramírez-Moreno
Instituto de Investigaciones en Materiales, Universidad Nacional Autónoma de México, Circuito exterior s/n, Ciudad Universitaria, Del. Coyoacán, México DF, Mexico and Departamento de Ingeniería en Metalurgia y Materiales, Escuela Superior de Ingeniería Química e Industrias Extractivas, IPN, UPALM, México DF, Mexico

Issis C. Romero-Ibarra
Instituto de Investigaciones en Materiales, Universidad Nacional Autónoma de México, Circuito exterior s/n, Ciudad Universitaria, Del. Coyoacán, México DF, Mexico

Juan Carlos Serrano-Ruiz
Advanced Materials Laboratory, Department of Inorganic Chemistry, University of Alicante, Apartado 99, Alicante, Spain

Shujuan Sui
MOE Key Laboratory of Bio-based Material Science and Technology, Northeast Forestry University, Harbin 150040, China

Jianping Sun
MOE Key Laboratory of Bio-based Material Science and Technology, Northeast Forestry University, Harbin 150040, China

Pascal Van Der Voort
Center for Ordered Materials, Organometallics & Catalysis (COMOC), Department of Inorganic and Physical Chemistry, Ghent University, Krijgslaan 281, Building S3, Ghent B-9000

Qingwen Wang
MOE Key Laboratory of Bio-based Material Science and Technology, Northeast Forestry University, Harbin 150040, China

Zhouyang Xiang
Departments of Chemistry and Forest Biomaterials, North Carolina State University, Raleigh, NC 27695-8005, USA

Rendang Yang
State Key Laboratory Pulp and Paper Engineering, South China University of Technology, Guangzhou 510460, China

Zhijun Zhang
MOE Key Laboratory of Bio-based Material Science and Technology, Northeast Forestry University, Harbin 150040, China and Department of Chemistry, Mississippi State University, Mississippi State, MS 39762, USA

Introduction

The search for sustainable energy resources is one of this century's great challenges. Biofuels (fuels produced from biomass) have emerged as one of the most promising renewable energy sources, offering the world a solution to its fossil-fuel addiction. They are sustainable, biodegradable, and contain fewer environmental contaminants than fossil fuels.

One of the biggest difficulties with biofuel production, however, is grabbing carbon for fuel while also removing oxygen from biomass. Unlike fossil fuel sources, biomass is rich in oxygen, which makes its transformation into fuel challenging. Catalysis offers a cheap, efficient, and sustainable way to remove oxygen from biomass, increasing its potential usefulness to the world's energy needs.

Heterogeneous catalysis has a long history of facilitating energy-efficient selective molecular transformations. It already contributes to most chemical manufacturing processes and to many industrial products. Catalysis also plays a central role in overcoming the barriers to economical and sustainable biofuel production.

Technical advances in catalyst design and processes are even more essential for second-generation biofuels produced from non-food feedstocks. As much oxygen as possible must be removed to gain the maximum energy density. In addition, the costs of oxygen removal need to be minimized.

Research that identifies inexpensive and efficient catalysts is crucial to the world's energy needs. Additionally, we need catalysts and catalytic processes that minimize hydrogen consumption, increase overall process activity, and gain high fuel yields. The research gathered in this compendium contributes to this vital field of investigation.

Juan Carlos Serrano-Ruiz, PhD

Concerns over the economics of proven fossil fuel reserves, in concert with government and public acceptance of the anthropogenic origin of rising CO_2 emissions and associated climate change from such combustible carbon, are driving academic and commercial research into new sustainable routes to fuel and chemicals. The quest for such sustainable resources to meet the demands of a rapidly rising global population represents one of this century's grand challenges. In Chapter 1, Lee discusses catalytic solutions to the clean synthesis of biodiesel, the most readily implemented and low cost, alternative source of transportation fuels, and oxygenated organic molecules for the manufacture of fine and speciality chemicals to meet future societal demands.

Chapter 2, by Bezergianni, argues that catalytic hydrotreatment of liquid biomass is a technology with the potential to overcome the limitations of biomass fuel productions. The author points to a wide range of new alternative fuels that are being developed using this technology, arguing that they are more useful than those developed using older methods, and points to catalytic hydrotreatment as the future of biofuels.

Chapter 3, by Murzin and Holmbom, describes some of the contemporary methods for the chemical analysis of biomass-derived chemicals. All available methods could not have been treated in this review, therefore the focus was mainly on chromatographic methods. A more comprehensive overview of analytical methods was published several years ago by one of the authors [31,33]. In the current work, detailed procedures were discussed for only a few cases as the emphasis was laid more on general approaches.

Concerns about diminishing fossil fuel reserves along with global warming effects caused by increasing levels of CO_2 in the atmosphere are driving society toward the search for new renewable sources of energy that can substitute for coal, natural gas and petroleum in the current energy system. Lignocellulosic biomass is abundant, and it has the potential to significantly displace petroleum in the production of fuels for the transportation sector. Ethanol, the main biomass-derived fuel used today, has benefited from production by a well-established technology and by partial compatibility with the current transportation infrastructure, leading to the domination of the world biofuel market. However, ethanol suffers from important limitations as a fuel (e.g., low energy density, high solubility in water) than can be overcome by designing strategies to convert non-edible lignocellulosic biomass into liquid hydrocarbon fuels (LHF) chemically similar to those currently used in internal combustion engines. Chapter 4, by Serrano-Ruiz and Dumesic, describes the main routes available to carry out such deep chemical transformation (e.g., gasification, pyrolysis, and aqueous-phase catalytic processing), with particular emphasis on those pathways involving aqueous-phase catalytic reactions. These latter catalytic routes achieve the required transformations in biomass-derived molecules with controlled chemistry and high yields, but require pretreatment/hydrolysis steps to overcome the recalcitrance of lignocellulose. To be economically viable, these aqueous-phase routes should be carried out with a small number of reactors and with minimum utilization of external fossil fuel-based hydrogen sources, as illustrated in the examples presented here.

Catalytic refining of bio-oil by reacting with olefin/alcohol over solid acids can convert bio-oil to oxygen-containing fuels. In Chapter 5, Zhang and colleagues studied the reactivities of groups of compounds typically present in bio-oil with 1-octene (or 1-butanol) at 120 °C/3 h over Dowex50WX2, Amberlyst15, Amberlyst36, silica sulfuric acid (SSA) and $Cs_{2.5}H_{0.5}PW_{12}O_{40}$ supported on K_{10} clay ($Cs_{2.5}/K_{10}$, 30 wt. %). These compounds include phenol, water, acetic acid, acetaldehyde, hydroxyacetone, D-glucose and 2-hydroxymethylfuran. Mechanisms for the overall conversions were proposed. Other olefins (1,7-octadiene, cyclohexene, and 2,4,4- trimethylpentene) and alcohols (iso-butanol) with different activities were also investigated. All the olefins and alcohols used were effective but produced varying product selectivities. A complex model bio-oil, synthesized by mixing all the above-stated model compounds, was refined under similar conditions to test the catalyst's activity. SSA shows the highest hydrothermal stability. $Cs_{2.5}/K_{10}$ lost most of its activity. A global reaction pathway is outlined. Simultaneous and competing esterification, etherfication, acetal formation, hydration, isomerization and other equilibria were involved. Synergistic interactions among reactants and products were determined. Acid-catalyzed olefin hydration removed water and drove the esterification and acetal formation equilibria toward ester and acetal products.

A newly designed downdraft wood stove achieved low-emission heating by integrating an alumina-supported mixed metal oxide catalyst in the combustion chamber operated under high temperature conditions. In the first step in Chapter 6, by Bindig and colleagues, a catalyst screening has been carried out with a lab-scale plug flow reactor in order to identify the potentially active mixed metal oxide catalysts. Mixed metal oxide catalysts have been the center of attention because of their expected high temperature stability and activity. The catalyst has been synthesized through two novel routes, and it has been integrated into a downdraft wood stove. The alumina-supported mixed metal oxide catalyst reduced the volatile hydrocarbons, carbon monoxide and carbonaceous aerosols by more than 60%.

As part of a programme aimed at exploiting lignin as a chemical feedstock for less oxygenated fine chemicals, several catalytic C–C bond forming reactions utilising guaiacol imidazole sulfonate are demonstrated in Chapter 7, by Leckie and colleagues. These include the cross-coupling of a Grignard, a non-toxic cyanide source, a benzoxazole, and nitromethane. A modified Meyers reaction is used to accomplish a second constructive deoxygenation on a benzoxazole functionalised anisole.

5-Halomethylfurfurals can be considered as platform chemicals of high reactivity making them useful for the preparation of a variety of important compounds. In Chapter 8, by Gao and colleagues, a one-pot route for the conversion of carbohydrates into 5-chloromethylfurfural (CMF) in a simple and efficient ($HCl-H_3PO_4/CHCl_3$) biphasic system has been investigated. Monosaccharides such as D-fructose, D-glucose and sorbose, disaccharides such as sucrose and cellobiose and polysaccharides such as cellulose were successfully converted into CMF in satisfactory yields under mild conditions. Our data shows that when using D-fructose the optimum yield of CMF was about 47%. This understanding allowed us to extent our work to biomaterials, such as wood powder and wood pulps with yields of CMF obtained being comparable to those seen with some of the enumerated mono and disaccharides. Overall, the proposed (HCl-H3PO4/CHCl3) optimized biphasic system provides a simple, mild, and cost-effective means to prepare CMF from renewable resources.

Chapter 9, by de Canck and colleagues, describes how a Periodic Mesoporous Organosilica (PMO) functionalized with sulfonic acid groups has been successfully synthesized via a sequence of post-synthetic modification steps of a trans-ethenylene bridged PMO material. The double bond is functionalized via a bromination and subsequent substitution obtaining a thiol functionality. This is followed by an oxidation towards a sulfonic acid group. After full characterization, the solid acid catalyst is used in the acetylation of glycerol. The catalytic reactivity and reusability of the sulfonic acid modified PMO material is investigated. The catalyst showed a catalytic activity and kinetics that are comparable with the commercially available resin, Amberlyst-15, and furthermore the catalyst can be recycled for several subsequent catalytic runs and retains its catalytic activity.

Chapter 10, by Ramírez-Morenoe et al., offers an alternative method to reduce CO_2 emissions through the use of alkaline and/or alkaline-earth oxide ceramics. These are able to selectively trap CO_2 under different conditions and suggests the feasibility of these kinds of solid for being used with different capture technologies and processes, such as: pressure swing adsorption (PSA), vacuum swing adsorption (VSP), temperature swing adsorption (TSA) and water gas shift reaction (WGSR). Therefore, the fundamental study regarding this matter can help to elucidate the whole phenomena in order to enhance the sorbents' properties.

PART I

OVERVIEW

CHAPTER 1

Catalysing Sustainable Fuel and Chemical Synthesis

ADAM F. LEE

1.1 INTRODUCTION

Sustainability, in essence the development of methodologies to meet the needs of the present without compromising those of future generations has become a watchword for modern society, with developed and developing nations and multinational corporations promoting international research programmes into sustainable food, energy, materials and even city planning. In the context of energy and materials (specifically synthetic chemicals), despite significant growth in proven and predicted fossil fuel reserves over the next two decades, notably heavy crude oil, tar sands, deepwater wells, and shale oil and gas, there are great uncertainties in the economics of their exploitation via current extraction methodologies,

Catalysing Sustainable Fuel and Chemical Synthesis. © *Lee AF.* Applied Petrochemical Research *4,1 (2014), doi: 10.1007/s13203-014-0056-z. Licensed under Creative Commons Attribution License, http://creativecommons.org/licenses/by/3.0.*

and crucially, an increasing proportion of such carbon resources (estimates vary between 65 and 80 % [1–3]) cannot be burned without breaching the UNFCC targets for a 2°C increase in mean global temperature relative to the pre-industrial level [4, 5]. There is clearly a tightrope to walk between meeting rising energy demands, predicted to rise 50 % globally by 2040 [6] and the requirement to mitigate current CO_2 emissions and hence climate change. The quest for sustainable resources to meet the demands of a rapidly rising global population represents one of this century's grand challenges [7, 8].

While many alternative sources of renewable energy have the potential to meet future energy demands for stationary power generation, biomass offers the most readily implemented, low cost solution to a drop-in transportation fuel for blending with/replacing conventional diesel [9] via carbohydrate hydrodeoxygenation (HDO) or lipid transesterification illustrated in Scheme 1. First generation bio-based fuels derived from edible plant materials received much criticism over the attendant competition between land usage for fuel crops versus traditional agricultural cultivation [10]. Deforestation practices, notably in Indonesia, wherein vast tracts of rainforest and peat land are being cleared to support palm oil plantations have also provoked controversy [11]. To be considered sustainable, second generation bio-based fuels and chemicals are sought that use biomass sourced from non-edible components of crops, such as stems, leaves and husks or cellulose from agricultural or forestry waste. Alternative non-food crops such as switchgrass or *Jatropha curcas* [12], which require minimal cultivation and do not compete with traditional arable land or drive deforestation, are other potential candidate biofuel feedstocks. There is also growing interest in extracting bio-oils from aquatic biomass, which can yield 80–180 times the annual volume of oil per hectare than that obtained from plants [13]. Approximately 9 % of transportation energy needs are predicted to be met via liquid bio-fuels by 2030 [14]. While the abundance of land and aquatic biomass, and particularly of agricultural, forestry and industrial waste, is driving the search for technologies to transform lignocellulose into fuels and chemical, energy and atom-efficient processes to isolate lignin and hemicellulose from the more tractable cellulose component, remain to be identified [15]. Thermal pyrolysis offers one avenue by which to obtain transportation fuels, and wherein catalysis will undoubt-

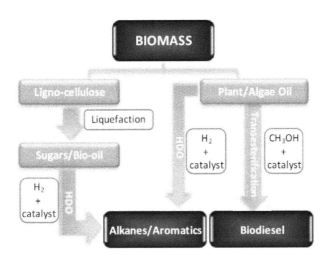

SCHEME 1: Chemical conversion routes for the co-production of chemicals and transportation fuels from biomass

edly play a significant role in both pyrolysis of raw biomass and subsequent upgrading of bio-oils via deoxygenation and carbon chain growth. Catalytic depolymerisation of lignin may also unlock opportunities for the production of phenolics and related aromatic compounds for fine chemical and pharmaceutical applications [16].

Biodiesel is a clean burning and biodegradable fuel which, when derived from non-food plant or algal oils or animal fats, is viewed as a viable alternative (or additive) to current petroleum-derived diesel [17]. Commercial biodiesel is currently synthesised via liquid base-catalysed transesterification of C_{14}–C_{20} triacylglyceride (TAG) components of lipids with C_1–C_2 alcohols [18–21] into fatty acid methyl esters (FAMEs) which constitute biodiesel as shown in Scheme 2, alongside glycerol as a potentially valuable by-product [22]. While the use of higher (e.g. C_4) alcohols is also possible [23], and advantageous in respect of producing a less polar and corrosive FAME [24] with reduced cloud and pour points [25], the current high cost of longer chain alcohols, and difficulties associated with separating the heavier FAME product from unreacted alcohol and glycerol, remain problematic. Unfortunately, homogeneous acid and base catalysts can corrode reactors and engine manifolds, and their removal from the resulting biofuel is particularly problematic and energy intensive, requiring aqueous quench and neutralisation steps which result in the formation of stable emulsions and soaps [9, 26, 27]. Such homogeneous approaches also yield the glycerine by-product, of significant potential value to the pharmaceutical and cosmetic industries, in a dilute aqueous phase contaminated by inorganic salts. Heterogeneous catalysis has a rich history of facilitating energy efficient selective molecular transformations and contributes to 90 % of chemical manufacturing processes and to more than 20 % of all industrial products [28, 29]. While catalysis has long played a pivotal role in petroleum refining and petrochemistry, in a post-petroleum era, it will face new challenges as an enabling technology to overcoming the engineering and scientific barriers to economically feasible routes to biofuels. The utility of solid base and acid catalysts for biodiesel production has been extensively reviewed [20, 30–33], wherein they offer improved process efficiency by eliminating the need for quenching steps, allowing continuous operation [34], and enhancing the purity of the glycerol by-product. Technical advances in catalyst and reactor design remain essential to utilise non-food based feedstocks and thereby ensure that biodiesel remains a key player in the renewable energy sector for the 21st century. Select pertinent developments in tailoring the nanostructure of solid acid and base catalysts for TAG transesterification to FAMEs and the related esterification of free fatty acid (FFAs) impurities common in bio-oil feedstocks are therefore discussed herein.

Biomass also offers the only non-fossil fuel route to organic molecules for the manufacture of bulk, fine and speciality chemicals and polymers [35] required to meet societal demands for advanced materials [8, 36]. The production of such highly functional molecules, whether derived from petroleum feedstocks, requires chemoselective transformations in which e.g. specific heteroatoms or functional groups are incorporated or removed without compromising the underpinning molecular properties. The selective oxidation (selox) of alcohols, carbohydrates and related α,β-unsaturated substrates represent an important class of reactions that

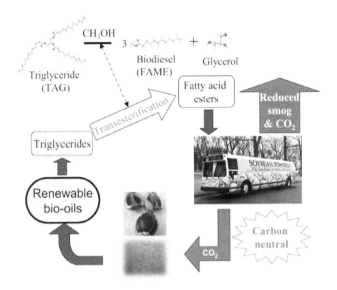

SCHEME 2: Carbon cycle for biodiesel production from renewable bio-oils via catalytic transesterification

underpin the synthesis of valuable chemical intermediates [37, 38]. The scientific, technological and commercial importance of green chemistry presents a significant challenge to traditional selox methods, which previously employed hazardous and toxic stoichiometric oxidants including permanganates, chromates and peroxides, with concomitant poor atom efficiencies and requiring energy-intensive separation steps to obtain the desired carbonyl or acid product. Alternative heterogeneous catalysts utilising oxygen or air as the oxidant offer vastly improved activity, selectivity and overall atom efficiency in alcohol selox (Scheme 3), but are particularly demanding due to the requirement to activate molecular oxygen and C–O bonds in close proximity at a surface in a solid–liquid–gas environment [39–41], and must also be scalable in terms of both catalyst synthesis and implementation. For example, continuous flow microreactors have been implemented in both homogeneous and heterogeneous aerobic selox, providing facile catalyst recovery from feedstreams for the latter [42, 43], but their scale-up/out requires complex manifolding to ensure adequate oxygen dissolution uniform reactant mixing and delivery [44, 45]. Efforts to overcome mass transport and solubility issues inherent to 3-phase catalysed oxidations have centred around the use of supercritical carbon dioxide to facilitate rapid diffusion of substrates to and products from the active catalyst site at modest temperatures [46] affording enhanced turnover frequencies (TOFs), selectivity and on-stream performance versus conventional batch operation in liquid organic solvents [47–51].

The past decade has seen significant progress in understanding the fundamental mode of action of Platinum Group Metal heterogeneous catalysts for aerobic selox and the associated reaction pathways and deactivation processes [41]. This insight has been aided by advances in analytical methodologies, notably the development of in situ or operando (under working conditions) spectroscopic [52–54] /microscopic [55–58] tools able to provide quantitative, spatio-temporal information on structure–function relations of solid catalysts in the liquid and vapour phase. Parallel improvements in inorganic synthetic protocols offer finer control over preparative methods to direct the nanostructure (composition, morphology, size, valence and support architecture) of palladium catalysts [59–61] and thereby enhance activity, selectivity and lifetime in an informed manner.

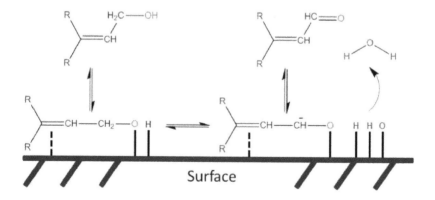

SCHEME 3: Cartoon depicting the atom-efficient, chemoselective aerobic selective oxidation of allylic alcohols to aldehydes over a heterogeneous catalyst

Ultimately, heterogeneous catalysts may offer significant advantages over homogeneous analogues in respect of initial catalyst cost, product separation, and metal recovery and recyclability [62]. Catalyst development can thus no longer be considered simply a matter of reaction kinetics, but as a clean technology wherein all aspects of process design, such as solvent selection, batch/flow operation, catalyst recovery and waste production and disposal are balanced [63]. The efficacy of Platinum Group Metals (PGMs) surfaces towards the liquid phase oxidation of alcohols has been known for over 50 years [64], and the development of heterogeneous platinum selox catalysts (and more recently coinage metals such as gold [65, 66]) the subject of recent reviews [39, 67–69] hence only palladium selox catalysis is described herein.

1.2 HETEROGENEOUSLY CATALYSED ROUTES TO BIODIESEL

1.2.1 SOLID ACID CATALYSED BIODIESEL SYNTHESIS

A wide range of inorganic and polymeric solid acids are commercially available, however, their application for the transesterification of oils into biodiesel has only been recently explored, in part reflecting their lower activity compared with base-catalysed routes [27], in turn necessitating higher reaction temperatures to deliver suitable conversions. While their activities are generally low, solid acids have the advantage that they are less sensitive to FFA contaminants than their solid base analogues, and hence can operate with unrefined feedstocks containing 3–6 wt% FFAs [27]. In contrast to solid bases which require feedstock pretreatment to remove fatty acid impurities, solid acids are able to esterify FFAs through to FAME in parallel with transesterification major TAG components without soap formation and thus reduce the number of processing steps to biodiesel [70–72].

Mesoporous silicas from the SBA family [73] have been examined for biodiesel synthesis, and include materials grafted with sulfonic acid groups [74, 75] or SO_4/ZrO_2 surface coatings [76]. Phenyl and propyl sulfonic acid SBA-15 catalysts are particularly attractive materials with activities comparable to Nafion and Amberlyst resins in palmitic acid esterification [77].

Phenylsulfonic acid functionalised silica is reportedly more active than their corresponding propyl analogues, in line with their respective acid strengths but is more difficult to prepare. Unfortunately, conventionally synthesised sulfonic acid functionalised SBA-15 silicas with pore sizes below ~6 nm possess long, isolated parallel channels and suffer correspondingly slow in-pore diffusion and catalytic turnover in FFA esterification. However, poragens such as trimethylbenzene [78] triethylbenzene or triisopropylbenzene [79] can induce swelling of the Pluronic P123 micelles used to produce SBA-15, enabling ordered mesoporous silicas with diameters spanning 5–30 nm, and indeed ultra-large-pores with a BJH pore diameter as much as 34 nm [79]. This methodology was recently applied to prepare a range of large pore SBA-15 materials employing trimethylbenzene as the poragen, resulting in the formation of highly ordered periodic mesostructures with pore diameters of ~6, 8 and 14 nm [80]. These silicas were subsequently functionalised by mercaptopropyl trimethoxysilane (MPTS) and oxidised with H_2O_2 to yield expanded $PrSO_3$-SBA-15 catalysts which were effective in both palmitic acid esterification with methanol and tricaprylin and triolein transesterification with methanol under mild conditions.

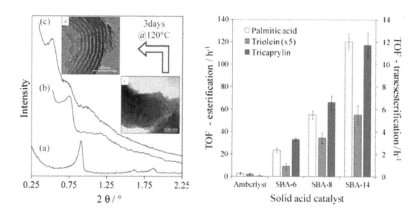

FIGURE 1: (Left) Low angle powder X-ray diffraction patterns and transmission electron micrographs of propylsulfonic acid functionalised SBA-15 silicas as a function of pore diameter; and (right) corresponding catalytic activity in FFA esterification and TAG transesterification compared to a commercial solid acid resin. Adapted from reference [80] with permission from The Royal Society of Chemistry

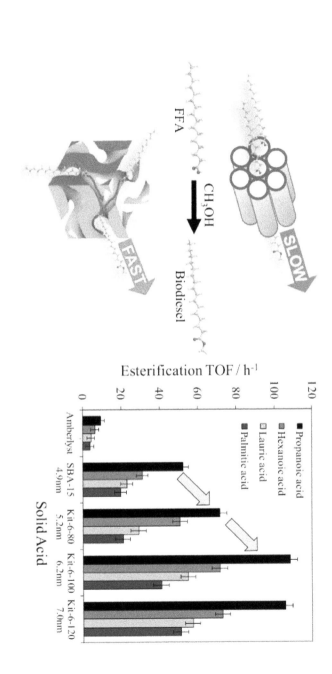

FIGURE 2: Superior performance of interconnected, mesoporous propylsulfonic acid KIT-6 catalysts for biodiesel synthesis via FFA esterification with methanol versus non-interconnected mesoporous SBA-15 analogue. Adapted from reference [82]. Copyright 2012 American Chemical Society

For both reactions, turnover frequencies dramatically increased with pore diameter, and all sulfonic acid heterogeneous catalysts significantly outperformed a commercial Amberlyst resin (Fig. 1). These rate enhancements are attributed to superior mass transport of the bulky FFA and triglycerides within the expanded $PrSO_3$-SBA-15. Similar observations have been made over Poly(styrenesulfonic acid)-functionalised ultra-large pore SBA-15 in the esterification of oleic acid with butanol [81].

Improving pore interconnectivity, for example through swapping the p6 *mm* architecture of SBA-15 for the *Ia3d* of KIT-6 was subsequently explored as an alternative means to enhance in-pore active site accessibility (Scheme 1) for FFA esterification [82]. KIT-6 mesoporous materials exhibit improved characteristics for biomolecule immobilisation [83] reflecting superior diffusion within the interconnected cubic structure. A family of pore-expanded propylsulfonic acid KIT-6 analogues were prepared via MPTS grafting and oxidation and screened for FFA esterification with methanol as a function of alkyl chain length under mild conditions. As-synthesised $PrSO_3$H-KIT-6 exhibited respective 40 and 70 % TOF enhancements toward propanoic and hexanoic acid esterification compared with a $PrSO_3$H-SBA-15 analogue of comparable (5 nm) pore diameter as a consequence of the improved mesopore interconnectivity. However, pore accessibility remained rate-limiting for esterification of the longer chain lauric and palmitic acids. Hydrothermal aging protocols facilitated expansion of the KIT-6 mesopore up to 7 nm, with consequent doubling of TOFs for lauric and palmitic acid esterification versus $PrSO_3$H-SBA-15 (Fig. 2).

While numerous solid acids have been applied for biodiesel synthesis [27, 32, 84], most materials exhibit micro- and/or mesoporosity which, as illustrated above, are not optimal for accommodating bulky C_{16}–C_{18} TAGs of FFAs. For example, incorporation of a secondary mesoporosity into a microporous H-β-zeolite to create a hierarchical solid acid significantly increased catalytic activity by lowering diffusion barriers [85]. Templated mesoporous materials are widely used as catalyst supports [86, 87], with SBA-15 silicas popular candidates for reactions pertinent to biodiesel synthesis as previously discussed [75, 77, 88]. However, such surfactant-templated supports possessing long, isolated parallel and narrow channels are ill-suited to efficient in-pore diffusion of bio-oil feedstocks affording poor catalytic turnover. Further improvements in pore architecture are hence required to optimise mass transport of heavier bulky TAGs and FFAs commonly found in plant and algal oils. Simulations demonstrate that in the Knudsen diffusion regime [89], where reactants/products are able to diffuse enter/exit mesopores but experience moderate diffusion limitations, hierarchical pore structures may significantly improve catalyst activity. Materials with interpenetrating, bimodal meso–macropore networks have been prepared using microemulsion [90] or co-surfactant [91] templating routes and are particularly attractive for liquid phase, flow reactors wherein rapid pore diffusion is required. Liquid crystalline (soft) and colloidal polystyrene nanospheres (hard) templating methods have been combined to create highly organised, macro–mesoporous aluminas [92] and 'SBA-15 like' silicas [93] (Scheme 4), in which both macro- and mesopore diameters can be independently tuned over the range 200–500 and 5–20 nm, respectively. The resulting hierarchical pore network of a propylsulfonic acid functionalised macro–mesoporous SBA-15 is shown in Fig. 3, wherein macropore incorporation confers a striking enhancement in the rates of tricaprylin transesterification and palmitic acid esterification with methanol, attributed to the macropores acting as transport conduits for reactants to rapidly access $PrSO_3H$ active sites located within the mesopores.

The hydrophilic nature of polar silica surfaces hinders their application for reactions involving apolar organic molecules. This is problematic for TAG transesterification (or FFA esterification) due to preferential in-pore diffusion and adsorption of alcohol versus fatty acid components. Surface hydroxyl groups also favour H_2O adsorption, which if formed during FFA esterification can favour the reverse hydrolysis reaction and consequent low FAME yields. Surface modification via the incorporation of organic functionality into polar oxide surfaces, or dehydroxylation, can lower their polarity and thereby increase initial rates of acid catalysed transformations of liquid phase organic molecules [94]. Surface polarity can also be tuned by incorporating alkyl/aromatic groups directly into the silica framework, for example polysilsesquioxanes can be prepared via the co-condensation of 1,4-bis(triethoxysilyl)benzene (BTEB), or 1,2-bis(trimethoxysilyl)-ethane (BTME), with TEOS and MPTS in the sol–gel process [95, 96] which enhances small molecule esterification [97] and etherification [98]. The incorporation of organic spectator groups (e.g. phenyl, methyl or propyl) during the sol–gel syntheses of SBA-15 [99] and MCM-41 [100] sulphonic acid silicas is achievable via co-grafting or simple addition of the respective alkyl or aryltrimethoxysilane during co-condensation protocols. An experimental and computational study of sulphonic acid functionalised MCM-41 materials was undertaken to evaluate the effect of acid site density and surface hydrophobicity on catalyst acidity and associated performance [101]. MCM-41 was an excellent candidate due to the availability of accurate models for the pore structure from kinetic Monte Carlo simulations [102], and was modified with surface groups to enable dynamic simulation of sulphonic acid and octyl groups co-attached within the MCM-41 pores. In parallel experiments, two catalyst series were investigated towards acetic acid esterification with butanol (Scheme 5). In one series, the propylsulphonic acid coverage was varied between θ (RSO$_3$H) = 0–100 % ML over the bare silica (MCM–SO$_3$H). For the second octyl co-grafted series, both sulfonic acid and octyl coverages were tuned (MCM–Oc–SO$_3$H). These materials allow the effect of lateral interactions between acid head groups and the role of hydrophobic octyl modifiers upon acid strength and activity to be separately probed.

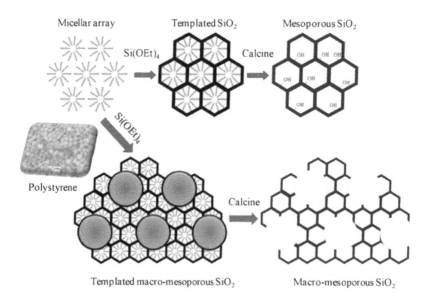

SCHEME 4: Liquid crystal and polystyrene nanosphere dual surfactant/physical templating route to hierarchical macroporous–mesoporous silicas

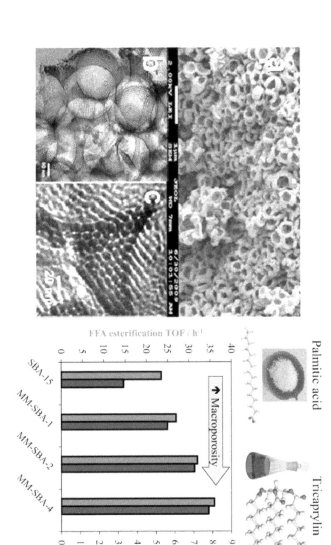

FIGURE 3: (Left) SEM (a) and low and high magnification TEM (b, c) micrographs of a hierarchical macro–mesoporous Pr-SO$_3$H-SBA-15; (right) corresponding catalytic performance in palmitic acid esterification and tricaprylin transesterification with methanol as a function of macropore density versus a purely mesoporous Pr-SO$_3$H-SBA-15. Adapted from reference [93] with permission from The Royal Society of Chemistry

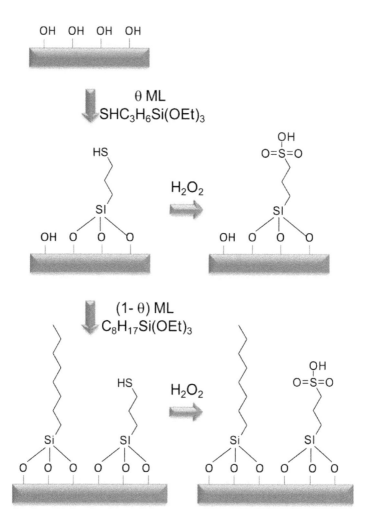

SCHEME 5: Protocol for the synthesis of sulfonic acid and octyl co-functionalised sulfonic acid MCM-41catalysts. Adapted from reference [101] with permission from The Royal Society of Chemistry

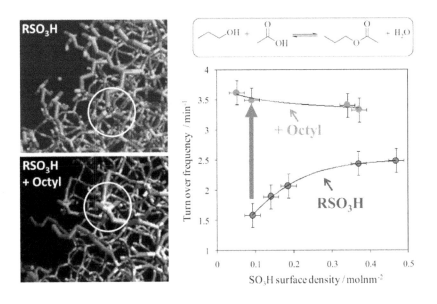

FIGURE 4: (Left) Molecular dynamics simulations of MCM–SO$_3$H and MCM–Oc–SO$_3$H pore models highlighting the interaction between surface sulfonic acid and hydroxyl groups in the absence of co-grafted octyl chains; (right) influence of PrSO$_3$H surface density and co-grafted octyl groups on catalytic performance in acetic acid esterification with butanol. Adapted from reference [101] with permission from The Royal Society of Chemistry

To avoid diffusion limitations, butanol esterification with acetic acid was selected as a model reaction (Fig. 4). Ammonia calorimetry revealed that the acid strength of polar MCM–SO$_3$H materials increases from 87 to 118 kJ mol^{-1} with sulphonic acid loading. Co-grafted octyl groups dramatically enhance the acid strength of MCM–Oc–SO$_3$H for submonolayer SO3H coverages, with _ΔHads(NH$_3$) rising to 10^3 kJ mol^{-1}. The per site activity of the MCM–SO$_3$H series in butanol esterification with acetic acid mirrors their acidity, increasing with SO$_3$H content. Octyl surface functionalisation promotes esterification for all MCM–Oc–SO$_3$H catalysts, doubling the turnover frequency of the lowest loading SO$_3$H material. Molecular dynamic simulations indicate that the interaction of isolated sulphonic acid moieties with surface silanol groups is the primary cause of the lower acidity and activity of submonolayer samples within the MCM–

SO$_3$H series. Lateral interactions with octyl groups help to re-orient sulphonic acid headgroups into the pore interior, thereby enhancing acid strength and associated esterification activity.

In summary, recent developments in tailoring the structure and surface functionality of sulfonic acid silicas have led to a new generation of tunable solid acid catalysts well-suited to the esterification of short and long chain FFAs, and transesterification of diverse TAGs, with methanol under mild reaction conditions. A remaining challenge is to extend the dimensions and types of pore-interconnectivities present within the host silica frameworks, and to find alternative low cost soft and hard templates to facilitate synthetic scale-up of these catalysts for multi-kg production. Surfactant template extraction is typically achieved via energy-intensive solvent reflux, which results in significant volumes of contaminated waste and long processing times, while colloidal templates often require high temperature calcination which prevents template recovery/re-use and releases carbon dioxide. Preliminary steps towards the former have been recently taken, employing room temperature ultrasonication in a small solvent volume to deliver effective extraction of the P123 Pluronic surfactant used in the preparation of SBA-15 in only 5 min, with a 99.9 % energy saving and 90 % solvent reduction over reflux methods, and without compromising textural, acidic or catalytic properties of the resultant Pr-SO$_3$H-SBA-15 in hexanoic acid esterification (Fig. 5) [103].

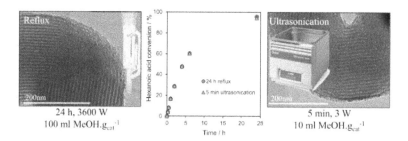

FIGURE 5: Surfactant template extraction via energy/atom-efficient ultrasonication delivers a one-pot PrSO$_3$H-SBA-15 solid acid catalyst with identical structure and reactivity to that obtained by conventional, inefficient reflux. Adapted from reference [103] with permission from The Royal Society of Chemistry

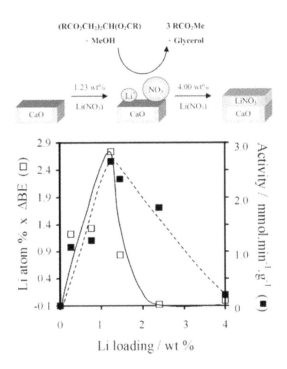

FIGURE 6: Correlation between evolving surface composition, density of electronically perturbed Li+ sites, and corresponding activity in tributyrin transesterification with methanol over Li-doped CaO as a function of Li loading. Adapted from reference [113] with permission from The Royal Society of Chemistry

1.3 SOLID BASE-CATALYSED BIODIESEL SYNTHESIS

Base catalysts are generally more active than acids in transesterification, and hence are particularly suitable for high purity oils with low FFA content. Biodiesel synthesis using a solid base catalyst in continuous flow, packed bed arrangement would facilitate both catalyst separation and co-production of high purity glycerol, thereby reducing production costs and enabling catalyst re-use. Diverse solid base catalysts are known, notably alkali or alkaline earth oxides, supported alkali metals, basic zeolites and clays such as hydrotalcites and immobilised organic bases [104]. Basicity in alkaline earth oxides is believed to arise from M^{2+}–O^{2-} ion pairs present in different coordination environments [105]. The strongest base sites occur at low coordination defect, corner and edge sites, or on high Miller index surfaces. Such classic heterogeneous base catalysts have been extensively tested for TAG transesterification [106] and there are numerous reports on commercial and microcrystalline CaO applied to rapeseed, sunflower or vegetable oil transesterification with methanol [107, 108]. Promising results have been obtained, with 97 % oil conversion achieved at 75 °C [108], however, concern remains over Ca^{2+} leaching under reaction conditions and associated homogeneous catalytic contributions [109], a common problem encountered in metal catalysed biodiesel production which hampers commercialisation [110].

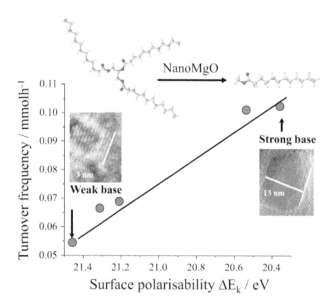

FIGURE 7: Relationship between surface polarisability of MgO nanocrystals and their turnover frequency towards tributyrin transesterifcation. Adapted from reference [117] with permission from The Royal Society of Chemistry

Alkali-doped CaO and MgO have also been investigated for TAG transesterification [111–113], with their enhanced basicity attributed to the genesis of O^- centres following the replacement of M^+ for M^{2+} and associated charge imbalance and concomitant defect generation. Optimum activity for Li-doped CaO occurs when a saturated Li^+ monolayer is formed (Fig. 6) [113], although leaching of the alkali promoter remains problematic [114].

It is widely accepted that the catalytic activity of alkaline earth oxide catalysts is very sensitive to their preparation, and corresponding surface morphology and/or defect density. For example, Parvulescu and Richards demonstrated the impact of the different MgO crystal facets upon the transesterification of sunflower oil by comparing nanoparticles [115] versus (111) terminated nanosheets [116]. Chemical titration reveals that both morphologies possess two types of base sites, with the nanosheets exhibiting well-defined, medium-strong basicity consistent with their uniform exposed facets and which confer higher FAMe yields during sunflower oil transesterification. Subsequent synthesis, screening and spectroscopic characterisation of a family of size-/shape-controlled MgO nanoparticles prepared via a hydrothermal synthesis revealed small (<8 nm) particles terminate in high coordination (100) facets, and exhibit both weak polarisability and poor activity in tributyrin transesterification with methanol [117]. Calcination drives restructuring and sintering to expose lower coordination stepped (111) and (110) surface planes, which are more polarisable and exhibit much higher transesterification activities under mild conditions. A direct correlation was therefore observed between the surface electronic structure and associated catalytic activity, revealing a pronounced structural preference for (110) and (111) facets (Fig. 7).

Hydrotalcites are another class of solid base catalysts that have attracted recent attention because of their high activity and robustness in the presence of water and FFA [118, 119]. Hydrotalcites ($[M(II)_{1-x} M(III)_x (OH)_2]^{x+} (A^n_{x/n}-) mH_2O$) adopt a layered double hydroxide structure with brucite-like $Mg(OH)_2$) hydroxide sheets containing octahedrally coordinated M^{2+} and M^{3+} cations and A^{n-} anions between layers to balance the overall charge [120], and are conventionally synthesised via co-precipitation from their nitrates using alkalis as both pH regulators and a carbonate source. Mg–Al hydrotalcites have been applied for TAG transesterification of poor and high quality oil feeds [121] such as refined and acidic cottonseed oil (9.5 wt% FFA), and animal fat feed (45 wt% water), delivering 99% conversion within 3 h at 200 °C. It is important to note that many catalytic studies employing hydrotalcites for transesterification are suspect due to their use of Na or K hydroxide/carbonate solutions to precipitate the hydrotalcite phase. Complete removal of alkali residues from the resulting hydrotalcites is inherently difficult, resulting in parallel ill-defined homogeneous contributions to catalysis arising from leached Na or K [122, 123]. This problem has been overcome by the development of alkali-free precipitation routes using NH_3OH and NH_3CO_3, offering well-defined thermally activated and rehydrated Mg–Al hydrotalcites with compositions spanning x = 0.25 − 0.55 [118]. Spectroscopic measurements reveal that increasing the Mg:Al ratio enables the surface charge and accompanying base strength to be systematically enhanced, with a concomitant increase in the rate of tributyrin transesterification under mild reaction conditions (Fig. 8).

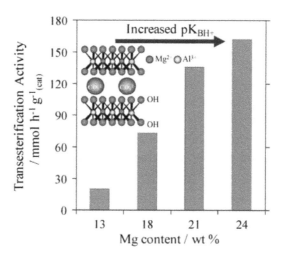

FIGURE 8: Impact of Mg:Al hydrotalcite surface basicity on their activity towards tributyrin transesterification. Adapted from reference [118] with permission from Elsevier

In spite of their promise for biodiesel production, conventionally prepared hydrotalcites are microporous, and hence poorly suited to application in the transesterification of bulky C_{16}–C_{18} TAG components of bio-oils. This problem was recently tackled by adopting the same hard templating method utilising polystyrene nanospheres described in Scheme 4 to incorporate macroporosity, and thus create a hierarchical macroporous–microporous hydrotalcite solid base catalyst [124]. The introduction of macropores as 'superhighways' to rapidly transport heavy TAG oil components to active base sites present at (high aspect ratio) hydrotalcite nanocrystallites, dramatically enhanced turnover frequencies for triolein transesterification compared with that achievable over an analogous Mg–Al microporous hydrotalcite (Fig. 9), reflecting superior mass transport through the hierarchical catalyst.

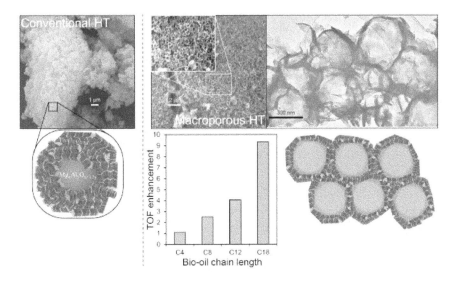

FIGURE 9: Superior catalytic performance of a hierarchical macroporous–microporous Mg–Al hydrotalcite solid base catalyst for TAG transesterification to biodiesel versus a conventional microporous analogue. Adapted from reference [124] with permission from The Royal Society of Chemistry

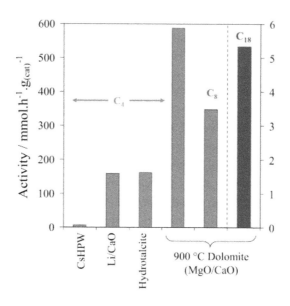

FIGURE 10: Catalytic activity of calcined Dolomite for the transesterification of short and long chain TAGs with methanol benchmarked against literature solid acid and base catalysts. Reproduced from reference [127] with permission from The Royal Society of Chemistry

In terms of sustainability, it is important to find low cost routes to the synthesis of solid base catalysts that employ earth abundant elements. Dolomitic rock, comprising alternating $Mg(CO_3)$–$Ca(CO_3)$ layers, is structurally very similar to calcite ($CaCO_3$), with a high natural abundance and low toxicity, and in the UK is sourced from quarries working Permian dolomites in Durham, South Yorkshire and Derbyshire [125]. In addition to uses in agriculture and construction, dolomite finds industrial applications in iron and steel production, glass manufacturing and as fillers in plastics, paints, rubbers, adhesives and sealants. Catalytic applications for powdered, dolomitic rock offer the potential to further valorise this readily available waste mineral, and indeed dolomite has shown promise in biomass gasification [126] as a cheap, disposable and naturally occurring material that significantly reduces the tar content of gaseous products from gasifiers. Dolomite has also been investigated as a solid base catalyst in biodiesel synthesis [127], wherein fresh dolomitic rock comprised approximately 77 % dolomite and 23 % magnesian calcite. High temperature calcination induced Mg surface segregation, resulting in MgO nanocrystals dispersed over $CaO/(OH)_2$ particles, while the attendant loss of CO_2 increases both the surface area and basicity. The resulting calcined dolomite proved an effective catalyst for the transesterification of C_4, C_8 and TAGs with methanol and longer chain C_{16-18} components present within olive oil, with TOFs for tributyrin conversion to methyl butanoate the highest reported for any solid base (Fig. 10). The slower transesterification rates for bulkier TAGs were attributed to diffusion limitations in their access to base sites. Calcined dolomite has also shown promise in the transesterification of canola oil with methanol, achieving 92 % FAME after 3 h reaction with 3 wt% catalyst [128].

In summary, a host of inorganic solid base catalysts have been developed for the low temperature transesterification of triglyceride components of bio-oil feedstocks, offering activities far superior to those achieved via alternative solid acid catalysts to date. However, leaching of alkali and alkaline earth elements and associated catalyst recycling remains a challenge, while improved resilience to water and fatty acid impurities in plant, algal and waste oils feedstocks is required to eliminate additional esterification pre-treatments. To date, only a handful of biodiesel production processes employing heterogeneous catalysts have been commercialised, notably the Esterfip-H process developed by Axens and IFP which utilises a mixture of ZnO and alumina and is operated on a 200 kton per annum scale with parallel production of high quality glycerine [129].

1.4 PALLADIUM CATALYSED AEROBIC ALCOHOL SELOX

1.4.1 PARTICLE SIZE EFFECTS

Within nanocatalysis, the particle size is a well-documented key parameter influencing both activity and selectivity. This reflects the combination of quantum and geometric effects associated with the respective evolution of electronic properties from atomic like to delocalised bands, and shifting population of low to high coordination surface atoms, with increasing nanoparticle size and dimensionality. Kaneda et al. [130] hypothesised that the unique reactivity of 2060 atom Pd clusters supported on titania towards aromatic alcohol selox arose from a distribution of Pd^0, Pd^+ and Pd^{2+} surfaces sites, with π-bonding interactions between the phenyl group and Pd^{2+} species facilitating subsequent oxidative addition of the O–H bond by neighbouring Pd^0 and eventual β–hydride elimination. Surface hydride was hypothesised to react with oxygen from a neighbouring Pd_2O centre forming H_2O and regenerating the metal site. Optimal activity for cinnamyl alcohol selox to cinnamaldehyde coincided with clusters possessing the maximum fraction of Pd^+ character.

Particle size dependency was also reported for the catalytic transformation of benzyl alcohol over Pd nanoparticles dispersed on alumina, SiO_2 and NaX zeolite supports [131, 132]. For Pd/NaX and Pd/SiO_2-Al_2O_3, benzyl alcohol selox was fastest over particles between 3 and 5 nm, whereas geraniol and 2-octanol were structure-insensitive. Systematic studies of particle size effects in cinnamyl and crotyl alcohol selox over amorphous and mesostructured alumina and silica supports have likewise uncovered pronounced size effects in both initial selox rates and TOFs [133–136], which increase monotonically with shrinking nanoparticle diameters (even down to single atoms) [137]. HAADF–STEM analysis reveals atomically dispersed palladium exhibits maximal rates towards benzyl, cinnamyl and crotyl alcohols, with selectivities to their corresponding aldehydes >70 %. The origin of such size effects is revisited below. The use of colloidal Pd nanoclusters for aqueous phase alcohol selox is limited [138–140], wherein Pd aggregation and Pd black formation hinders catalytic performance. However, the successful stabilisation of 3.6 nm Pd nanoclusters is reported using an amphiphilic nonionic triblock copolymer, Pluronic P123; in the selective oxidation of benzyl alcohol, 100 % aldehyde selectivity and high selox rates are achievable, with high catalytic activity maintained with negligible sintering after 13 recycling reactions [141].

1.4.2 SURFACE REACTION MECHANISM

The rational design and optimisation of palladium selox catalysts require a microscopic understanding of the active catalytic species responsible for alcohol and oxygen activation, and the associated reaction pathway to the aldehyde/ketone products and any competing processes. A key characteristic of palladium is its ability to perform selox chemistry at temperatures between 60 and 160 °C and with ambient oxygen pressure [39, 142] via the widely accepted oxidative dehydrogenation route illustrated in Scheme 3 [39, 67]. Whether O–H or C–H scission of the α-carbon is the first chemical step remains a matter of debate, since the only fundamental studies over well-defined Pd(111) surfaces to date employed temperature-programmed XPS [143] and metastable de-excitation spectroscopy (MDS) [144] with temporal resolutions on the second $\rightarrow$ minute timescale, over which loss of both hydrogens appears coincident. However, temperature-programmed mass spectrometric [145] and vibrational [146] studies of unsaturated C_1–C_3 alcohols implicate O–H cleavage and attendant alkoxy formation over Pd single crystal surfaces as the first reaction step [142, 147]. It is generally held that the resultant hydrogen adatoms react with dissociatively absorbed oxygen to form water, which immediately desorbs at ambient temperature thereby shifting the equilibrium to carbonyl formation [39, 67]. Temperature-programmed XPS studies of crotyl alcohol adsorbed over clean Pd(111) [143] prove that oxidative dehydrogenation to crotonaldehyde occurs at temperatures as low as −60 °C (Fig. 11), with alcohol dehydration to butane only a minor pathway. These ultra-high vacuum measurements also revealed that reactively formed crotonaldehyde undergoes a competing decarbonylation reaction over metallic palladium above 0°C yielding strongly bound CO and propylidene which may act as site-blockers poisoning subsequent catalytic selox cycles, coincident with evolution of propene into the gas phase. Unexpectedly, pre-adsorbed atomic oxygen switched-off undesired decarbonylation chemistry, promoting facile crotonaldehyde desorption.

1.4.3 NATURE OF THE ACTIVE SITE

The preceding observation that surface oxygen is not only critical for the removal of hydrogen adatoms but also to suppress decarbonylation of selox products over metallic palladium is in excellent agreement with an in situ ATR-IR study of cinnamyl alcohol selox over Pd/Al$_2$O$_3$ [148].

In related earlier investigations employing aqueous electrochemical protocols, the same researchers postulated that oxidative dehydrogenation of alcohols requires PGM catalysts in a reduced state, hypothesising that 'over-oxidation' was responsible for deactivation of palladium selox catalysts [69]. A subsequent operando X-ray absorption spectroscopy (XAS) study by Grunwaldt et al. [150], bearing remarkable similarity to an earlier study to the author of this review [149], evidenced in situ reduction of oxidised palladium in an as-prepared Pd/Al_2O_3 catalyst during cinnamyl alcohol oxidation within a continuous flow fixed-bed reactor. Unfortunately the reaction kinetics were not measured in parallel to explore the impact of palladium reduction, however, a follow-up study of 1-phenylethanol selox employing the same reactor configuration (and oxygen-deficient conditions) evidenced a strong interplay between selox conversion/selectivity and palladium oxidation state [151].

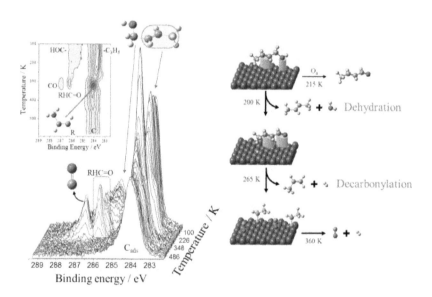

FIGURE 11: Temperature-programmed C 1s XP spectra of a reacting crotyl alcohol adlayer over Pd(111) highlighting the primary dehydrogenation pathway and competing decarbonylation pathways. Adapted from reference [143]. Copyright 2007 American Chemical Society

It was concluded that metallic Pd was the catalytically active species, an assertion re-affirmed in subsequent in situ ATR-IR/XAS measurements of benzyl [152–154] and cinnamyl alcohol [155] selox in toluene and under supercritical CO_2, respectively, wherein the C=O stretching intensity was assumed to track alcohol conversion. It is interesting to note that the introduction of oxygen to the reactant feed in these infrared studies dramatically improved alcohol conversion/aldehyde production (Fig. 12), which was attributed to hydrogen abstraction from the catalyst surface [156, 157] rather than to a change in palladium oxidation state. In contrast to their liquid phase experiments, high pressure XANES and EXAFS measurements of Pd/Al_2O_3 catalysed benzyl alcohol selox under supercritical CO_2 led Grunwaldt and Baiker to conclude that maximum activity arose from particles mainly oxidised in the surface/shelfedge [48].

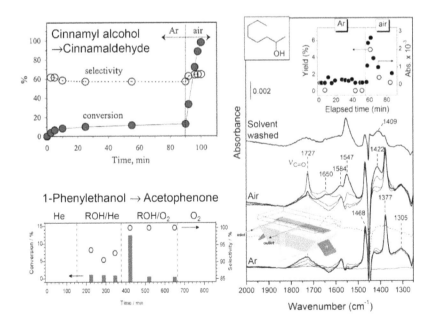

FIGURE 12: Impact of oxygen on the selective oxidation of (top left) cinnamyl alcohol; (bottom left) 1-phenylethanol; and (right) 2-octanol. Adapted from references [148, 151, 154] with permission from Elsevier

In a parallel research programme, the author's group systematically characterised the physicochemical properties of palladium nanoparticles as a function of size over non-reducible supports to quantify structure–function relations in allylic alcohol selox [133–137, 158, 159]. The combination of XPS and XAS measurements revealed that freshly prepared alumina [134, 137] and silica [135, 158] supported nanoparticles are prone to oxidation as their diameter falls below ~4 nm, with the fraction of PdO proportional to the support surface area and interconnectivity.

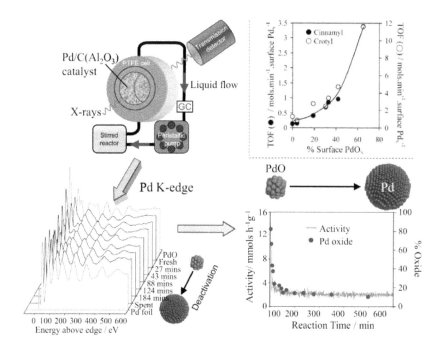

FIGURE 13: (Top right) Dependence of allylic alcohol selox rate upon surface PdO; (top left) schematic of operando liquid phase reactor; (bottom left) evolution of Pd K-edge XAS of Pd/Al$_2$O$_3$ catalyst during cinnamyl alcohol aerobic selox; (bottom right) temporal correspondence between Pd oxidation state and selox activity in cinnamyl alcohol selox. Adapted from references [133, 134] with permission from The Royal Society of Chemistry

Complementary kinetic analyses uncovered a direct correlation between the surface PdO content and activity/TOFs towards cinnamyl and crotyl alcohol selox [134, 137]. Operando liquid phase XAS of Pd/C and Pd/Al_2O_3-SBA-15 catalysts during cinnamyl alcohol selox evidenced in situ reduction of PdO (Fig. 13), however, by virtue of simultaneously measuring the rate of alcohol selox, Lee et al. were able to prove that this oxide → metal structural transition was accompanied by coincident deactivation. Together these findings strongly implicate a (surface) PdO active phase, consistent with surface science predictions that metallic palladium favours aldehyde decarbonylation and consequent self-poisoning by CO and organic residues [143, 160], akin to that reported during fatty acid decarboxylation over Pd/MCF [161].

To conclusively establish whether oxide or metal is responsible for alcohol selox catalysed by dispersed palladium nanoparticles, a multidimensional spectroscopic investigation of vapour phase crotyl alcohol selox was undertaken (since XAS is an averaging technique a complete understanding of catalyst operation requires multiple analytical techniques [162–164]). Synchronous, time-resolved DRIFTS/MS/XAS measurements of supported and colloidal palladium were performed in a bespoke environmental cell [165] to simultaneously interrogate adsorbates on the catalyst surface, Pd oxidation state and reactivity under transient conditions in the absence of competitive solvent effects [166, 167]. Under mild reaction temperatures, palladium nanoparticles were partially oxidised, and unperturbed by exposure to sequential alcohol or oxygen pulses (Fig. 14). Crotonaldehyde formed immediately upon contact of crotyl alcohol with the oxide surface, but only desorbed upon oxygen co-adsorption. Higher reaction temperatures induced PdO reduction in response to crotyl alcohol exposure, mirroring that observed during liquid phase selox, however, this reduction could be fully reversed by subsequent oxygen exposure. Such reactant-induced restructuring was exhibited by all palladium nanoparticles, but the magnitude was inversely proportional to particle size [168]. These dynamic measurements decoupled the relative reactivity of palladium oxide from metal revealing that PdO favoured crotyl alcohol selox to crotonaldehyde and crotonic acid, whereas metallic palladium drove secondary decarbonylation to propene and CO in accordance with surface science predictions [143].

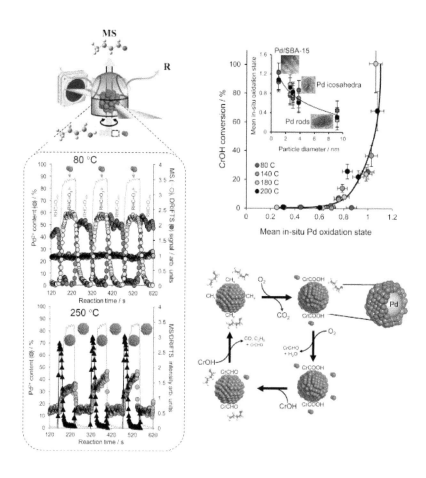

FIGURE 14: (Left) Cartoon of operando DRIFTS/MS/XAS reaction cell and resulting temperature dependent behaviour of Pd oxidation state and associated reactivity towards crotyl alcohol oxidation over a Pd/meso-Al$_2$O$_3$ catalyst—only selective oxidation over surface PdO occurs at 80 °C, whereas crotonaldehyde decarbonylation and combustion dominate over Pd metal at 250 °C; (top right) relationship between Pd oxidation derived in situ and crotyl alcohol conversion; (bottom right) summary of reaction-induced redox processes in Pd-catalysed crotyl alcohol selox. Adapted with permission from references [166, 168]. Copyright 2011 and 2012 American Chemical Society

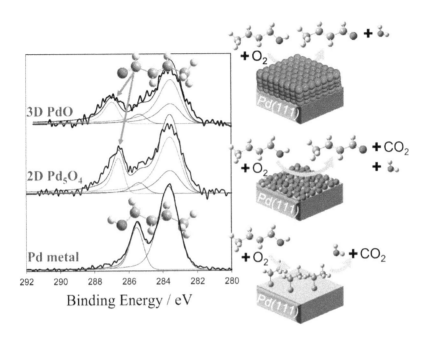

FIGURE 15: (Left) C 1s XP spectra of crotyl alcohol/O_2 gas mixture over metallic and oxidised Pd(111) surfaces; (right) differing reactivity of palladium metal and oxide surfaces. Adapted from reference [169]. Copyright 2012 American Chemical Society

Recent ambient pressure XPS investigations of crotyl alcohol/O_2 gas mixtures over metallic and oxidised Pd(111) single crystal surfaces confirmed that only two-dimensional Pd_5O_4 and three-dimensional PdOx surfaces were capable of crotonaldehyde production (Fig. 15) [169]. However, even under oxygen-rich conditions, on-stream reduction of the Pd_5O_4 monolayer oxide occurred >70 °C accompanied by surface poisoning by hydrocarbon residues. In contrast, PdOx multilayers were capable of sustained catalytic turnover of crotyl alcohol to crotonaldehyde, conclusively proving surface palladium oxide as the active phase in allylic alcohol selox.

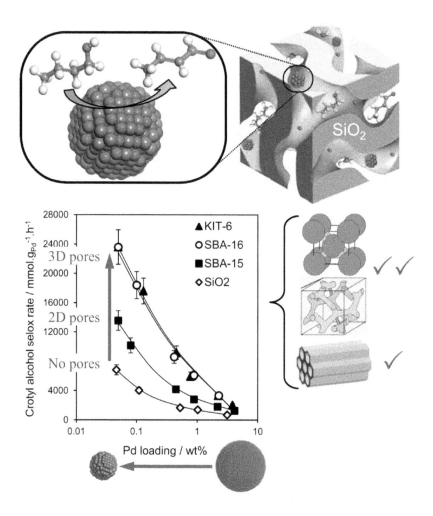

FIGURE 16: Comparative activity of Pd nanoparticles dispersed over amorphous, 2D non-interconnected SBA-15 and 3D interconnected SBA-16 and KIT-6 mesoporous silicas in the selective aerobic oxidation of crotyl alcohol. Adapted from reference [135]. Copyright 2011 American Chemical Society

1.4.4 ESTABLISHING SUPPORT EFFECTS

Anchoring Pd nanoparticles onto support structures offers an effective means to tune their physicochemical characteristics and prevent on-stream deactivation e.g. by sintering. Supports employing porous architectures, acid/base character and/or surface redox chemistry e.g. strong metal support interaction (SMSI), afford further opportunities to influence catalyst reactivity [170–173]. Mesoporous silicas are widely used to disperse metal nanoparticles [135, 136, 171, 174, 175]. The transition from low surface area, amorphous silica (200 m^{2}g^{-1}) to two-dimensional non-interconnected pore channels (SBA-15) [73] and three-dimensional interconnected porous frameworks (SBA-16, KIT-6) [73, 176, 177] improved the dispersion of Pd nanoparticles and hence degree of surface oxidation and thus activity in allylic alcohol selox (Fig. 16), but had little impact on the mass transport of small alcohols to/from the active site. [135, 136] The high thermal and chemical stability of such mesoporous silica [178, 179] makes such supports well-suited to commercialisation. Pd nanoparticles confined within such mesoporous silicas demonstrate good selectivity in crotyl and cinnamyl alcohol selox to their respective aldehydes (>70 %), and excellent TOFs of 7,000 and 5,000 h^{-1} for the respective alcohols. Similar activities are reported for secondary and tertiary allylic alcohols, highlighting the versatility of silica supported Pd nanoparticles [51, 135, 136, 180–182]. Incorporation of macropores into SBA-15 via dual hard/soft templating to form a hierarchically ordered macroporous–mesoporous Pd/SBA-15 was recently shown to promote the catalytic selox of sterically challenging sesquiterpenoid substrates such as farnesol and phytol via (1) stabilising PdO nanoparticles and (2) dramatically improving in-pore diffusion and access to active sites [158].

The benefits of mesostructured supports are not limited to silica, with ultra-low loadings of palladium impregnated onto a surfactant-templated mesoporous alumina (350 m^2 g^{-1}) generating atomically dispersed Pd^{2+} centres [137]. Such single-site catalysts were 10 times more active in crotonaldehyde and cinnamaldehyde production than comparable materials employing conventional (100 m^2 g^{-1}) γ-alumina, owing to the preferential genesis of higher concentrations of electron-deficient palladium [134, 137], due to either pinning at cation vacancies or metal → support charge transfer [183]. These Pd/meso-Al$_2$O$_3$ catalysts exhibited similar TOFs to their silica counterparts (7,080 and 4,400 h^{-1} for crotyl and cinnamyl alcohol selox, respectively) [137], consistent with a common active site and reaction mechanism (Fig. 17).

Mesoporous titania and ceria have also attracted interest as novel catalyst supports. The oxygen storage capacity of ceria-derived materials is of particular interest due to their facile Ce$^{3+}\leftrightarrow$Ce^{4+} redox chemistry [173, 184–188]. Sacrificial reduction of the ceria supports by reactively formed hydrogen liberated during the oxidative dehydrogenation of alcohols could mitigate in situ reduction of oxidised palladium, and hence maintain selox activity and catalyst lifetime, with Ce^{4+} sites regenerated by dissociatively adsorbed gas phase oxygen [187, 189, 190]. Due to its high density, conventional nanocrystalline cerias possess meagre surface areas (typically ~5 m^{2}g^{-1}), hence Pd/CeO$_2$ typically exhibit poor selox behaviour due to their resultant low nanoparticle dispersions which favour (self-poisoning) metallic Pd [189, 191, 192].

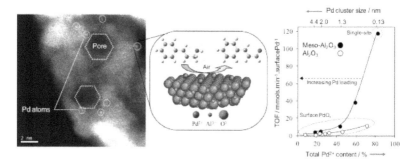

FIGURE 17: (Left) HAADF–STEM image of atomically dispersed Pd atoms on a mesoporous Al$_2$O$_3$ support; and (right) associated relationship between Pd^{2+} content/ dispersion and activity in crotyl alcohol selox over Pd/alumina catalysts. Adapted with permission from reference [137]. Copyright Wiley–VCH Verlag GmbH & Co. KGaA

1.4.5 BIMETALLIC PALLADIUM SELOX CATALYSTS

Incorporation of a second metal into palladium catalysts can improve both alcohol selox stability and selectivity. Typical promoters such as Ag, Bi, Pb and Sn [157, 193–196], enhance oxidation performance towards challenging substrates such as propylene glycol [197] as well as allylic and benzylic alcohols. Wenkin et al. [194] reported glucose oxidation to gluconates was increased by a factor of 20 over Pd–Bi/C catalysts (Bi/Pds = 0.1) versus Pd/C counterparts. In situ XAS and attenuated total reflection infrared spectroscopy (ATR-IR) suggested that Bi residing at the catalyst surface protects palladium from deactivation by either over-oxidation (a hypothesis since disproved [166, 167, 169]) or site-blocking by aromatic solvents [153]. Prati et al. [200] first reported significant rate enhancements and resistance to deactivation phenomena in the liquid phase selox of d-sorbitol to gluconic/gulonic acids upon addition of Au to Pd/C and Pt/C materials [198], subsequently extended to polyol and long chain aliphatic alcohols [199]. A strong synergy between Pd and Au centres was also demonstrated by Hutchings et al., wherein Au–Pd alloy nanoparticles supported on titania exhibited increased reactivity towards a diverse range of primary, allylic and benzylic alkyl alcohols compared to monometallic palladium analogues. The versatility of Au–Pd catalysts has also been shown in selox of saturated hydrocarbons [201], ethylene glycol [202], glycerol [203] and methanol [204], wherein high selectivity and resistance to on-stream deactivation is noted.

The effect of Au–Pd composition has been extensively studied for bimetallic nanoparticles stabilised by PVP surfactants [205]. An optimal Au:Pd composition of 1:3 was identified for 3 nm particles towards the aqueous phase aerobic selox of benzyl alcohol, 1-butanol, 2-butanol, 2-buten-1-ol and 1,4-butanediol; in each case the bimetallic catalysts were superior to palladium alone. Mertens et al. [206] examined similar systems utilising 1.9 nm nanoparticles, wherein an optimal Au content of around 80 % was determined for benzyl alcohol selox. The synergic interaction between Au and Pd therefore appears interdependent on nanoparticle size. It is well-known that the catalytic activity of Au nanoparticles increases dramatically <2 nm [207], hence it is interesting to systematically compare phase separated and alloyed catalysts. The author's group prepared titania-supported Au shell (5-layer)-Pd core (20 nm) bimetallic nanoparticles for the liquid phase selox of crotyl alcohol and systematically studied the evolution of their bulk and surface properties as a function of thermal processing by in situ XPS, DRIFTS, EXAFS, XRD and ex-situ HRTEM. Limited Au/Pd alloying occurred below 300 °C in the absence of particle sintering [208]. Higher temperatures induced bulk and surface alloying, with concomitant sintering and surface roughening. Migration of Pd atoms from the core to the surface dramatically enhanced activity and selectivity, with the most active and selective surface alloy containing 40 atom % Au (Fig. 18). This discovery was rationalised in terms of complementary temperature-programmed mass spectometric studies of crotyl alcohol and reactively formed intermediates over Au/Pd(111) model single crystal catalysts which reveal that gold–palladium alloys promote desorption of the desired crotonaldehyde selox product while co-adsorbed oxygen adatoms actually suppress aldehyde combustion. In contrast, the combustion of propene, the undesired secondary product of crotonaldehyde decarbonylation, is enhanced by co-adsorbed oxygen [160].

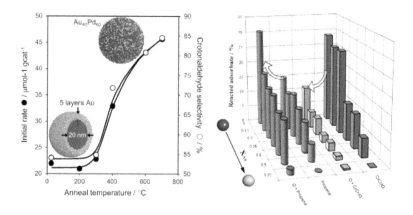

FIGURE 18: Impact of thermally induced Au–Pd alloying of (left) titania-supported Au shell–Pd core nanoparticles on crotyl alcohol aerobic selox adapted from reference [208], with permission from Elsevier; and (right) ultrathin gold overlayers on Pd(111) on crotonaldehyde and propene decomposition with/without co-adsorbed oxygen, adapted from reference [160] with permission from the PCCP Owner Societies

Scott et al. prepared the inverse Au core-Pd shell nanoparticles and explored the catalytic cycle for alcohol selox to assess their associated stability [205, 209–212]. In situ Pd–K and Pd–LIII edge XAS of a Au nanoparticle/Pd(II) salt solution were undertaken to discriminate two possible reaction mechanisms. No evidence was found that crotyl alcohol oxidation was accompanied by Pd^{2+} reduction onto Au nanoparticles, resulting in the formation of a metallic Pd shell (with oxygen subsequently regenerating electron-deficient palladium), and therefore proposed β-H elimination as the favoured pathway. Scott and co-workers proposed that the Au core prevents the re-oxidation of surface Pd0 atoms; no Pd–O and Pd–Cl contributions were observed by EXAFS.

In summary, the selective oxidation of complex alcohol substrates can be accomplished through Pd-mediated heterogeneous catalysis with high turnover and product selectivity. Application of in situ and operando techniques, such as X-ray and IR spectroscopies, has elucidated the mechanism of alcohol oxidative dehydrogenation and competing aldehyde decarbonylation. Surface PdO has been identified as the active catalytic species, and deactivation the result of reduction to metallic palladium and concomitant self-poisoning by strongly bound CO and carbonaceous residues. Breakthroughs in analytical tools and synthetic approaches to engineering nanoporous supports and shape/size controlled nanoparticles have delivered significant progress towards improved atom and energy efficiency and catalyst stability, however, next generation palladium selox catalysts necessitate improved synthetic protocols to create higher densities of ultra-dispersed Pd^{2+} centres with superior resistance to on-stream reduction under atmospheric oxygen.

REFERENCES

1. S. Kretzmann, http://priceofoil.org/2013/11/26/new-analysis-shows-growing-fossil-reserves-shrinking-carbon-budget/
2. C.C. Authority, Reducing Australia's greenhouse gas emissions—targets and progress review draft report, commonwealth of Australia, 2013
3. C.C. Secretariat, The critical decade 2013 climate change science, risks and responses, commonwealth of Australia, 2013

4. I.E. Agency, prospect of limiting the global increase in temperature to 2 °C is getting bleaker, http://www.iea.org/newsroomandevents/news/2011/may/name,19839,en.html
5. I.E. Agency, Redrawing the energy climate map, 2013
6. U.S.E.I. Administration, international energy outlook 2013, 2013
7. Armaroli N, Balzani V (2007) Angew Chem Int Ed 46:52–66
8. Azadi P, Inderwildi OR, Farnood R, King DA (2013) Renew Sustain Energy Rev 21:506–523
9. Demirbas A (2007) Energy Policy 35:4661–4670
10. McLaughlin DW (2011) Conserv Biol 25:1117–1120
11. Danielsen F, Beukema H, Burgess ND, Parish F, BrÜHl CA, Donald PF, Murdiyarso D, Phalan BEN, Reijnders L, Struebig M, Fitzherbert EB (2009) Conserv Biol 23:348–358
12. Achten WMJ, Verchot L, Franken YJ, Mathijs E, Singh VP, Aerts R, Muys B (2008) Biomass Bioenergy 32:1063–1084
13. Mata TM, Martins AA, Caetano NS (2010) Renew Sustain Energy Rev 14:217–232
14. BP, BP energy outlook 2030, 2011
15. Sheldon RA (2014) Green Chem 16:950–963
16. Pandey MP, Kim CS (2011) Chem Eng Technol 34:29–41
17. Knothe G (2010) Top Catal 53:714–720
18. Climent MJ, Corma A, Iborra S, Velty A (2004) J Catal 221:474–482
19. Constantino U, Marmottini F, Nocchetti M, Vivani R (1998) Eur J Inorg Chem 10:1439–1446
20. Narasimharao K, Lee A, Wilson K (2007) J Biobased Mater Bioenergy 1:19–30
21. Othman MR, Helwani Z, Martunus, Fernando WJN (2009) Appl Organomet Chem 23:335–346
22. Liu Y, Lotero E, Goodwin JG, Mo X (2007) Appl Catal A 33:138–148
23. Geuens J, Kremsner JM, Nebel BA, Schober S, Dommisse RA, Mittelbach M, Tavernier S, Kappe CO, Maes BUW (2007) Energy Fuels 22:643–645
24. Hu, Du, Tang Z, Min (2004) Ind Eng Chem Res 43:7928–7931
25. Knothe G (2005) Fuel Process Technol 86:1059–1070
26. Ma F, Hanna MA (1999) Bioresour Technol 70:1–15.
27. Lotero E, Liu Y, Lopez DE, Suwannakarn K, Bruce DA, Goodwin JG (2005) Ind Eng Chem Res 44:5353–5363
28. Thomas JM (2012) Proc R Soc A Math Phys Eng Sci 468:1884–1903
29. Somorjai GA, Frei H, Park JY (2009) J Am Chem Soc 131:16589–16605
30. Luque R, Herrero-Davila L, Campelo JM, Clark JH, Hidalgo JM, Luna D, Marinas JM, Romero AA (2008) Energy Environ Sci 1:542–564
31. Luque R, Lovett JC, Datta B, Clancy J, Campelo JM, Romero AA (2010) Energy Environ Sci 3:1706–1721
32. Dacquin J-P, Lee AF, Wilson K (2010) Thermochemical conversion of biomass to liquid fuels and chemicals. The Royal Society of Chemistry, UK, pp 416–434
33. Wilson K, Lee AF (2012) Catal Sci Technol 2:884–897
34. Eze VC, Phan AN, Pirez C, Harvey AP, Lee AF, Wilson K (2013) Catal Sci Technol 3:2373–2379
35. Chen G-Q, Patel MK (2011) Chem Rev 112:2082–2099

36. Bozell JJ, Petersen GR (2010) Green Chem 12:539–554
37. Sheldon RA (2005) Green Chem 7:267–278
38. Sheldon RA, Arends I, Hanefeld U (2007) Green chemistry and catalysis. Wiley-VCH Verlag GmbH & Co. KGaA, Weinheim
39. Mallat T, Baiker A (2004) Chem Rev 104:3037–3058
40. Lee AF (2014). In: Wilson K, Lee AF (eds) Heterogeneous catalysts for clean technology: spectroscopy, design, and monitoring chapt. 2. Wiley-VCH Verlag GmbH & Co. KGaA, Weinheim, pp 11–33
41. Vinod CP, Wilson K, Lee AF (2011) J Chem Technol Biotechnol 86:161–171
42. Kobayashi S, Miyamura H, Akiyama R, Ishida T (2005) J Am Chem Soc 127:9251–9254
43. Kaizuka K, Lee KY, Miyamura H, Kobayashi S (2012) J Flow Chem 2:1
44. Ye XA, Johnson MD, Diao TN, Yates MH, Stahl SS (2010) Green Chem 12:1180–1186
45. Ayude A, Cechini J, Cassanello M, Martinez O, Haure P (2008) Chem Eng Sci 63:4969–4973
46. Tschan R, Wandeler R, Schneider MS, Schubert MM, Baiker A (2001) J Catal 204:219–229
47. Caravati M, Grunwaldt JD, Baiker A (2004) Catal Today 91–2:1–5
48. Grunwaldt JD, Caravati M, Baiker A (2006) J Phys Chem B 110:9916–9922
49. Burgener M, Tyszewski T, Ferri D, Mallat T, Baiker A (2006) Appl Catal A Gen 299:66–72
50. Hou ZS, Theyssen N, Leitner W (2007) Green Chem 9:127–132
51. Hou Z, Theyssen N, Brinkmann A, Klementiev KV, Grünert W, Bühl M, Schmidt W, Spliethoff B, Tesche B, Weidenthaler C (2008) J Catal 258:315–323
52. Beale AM, Jacques SDM, Weckhuysen BM (2010) Chem Soc Rev 39:4656–4672
53. Grunwaldt JD, Schroer CG (2010) Chem Soc Rev 39:4741–4753
54. Lee AF (2012) Aust J Chem 65:615–623
55. Gai PL, Sharma R, Ross FM (2008) MRS Bull 33:107–114
56. Jungjohann KL, Evans JE, Aguiar JA, Arslan I, Browning ND (2012) Microsc Microanal 18:621–627
57. Browning ND, Bonds MA, Campbell GH, Evans JE, Lagrange T, Jungjohann KL, Masiel DJ, Mckeown J, Mehraeen S, Reed BW, Santala M (2012) Curr Opin Solid State Mat Sci 16:23–30
58. Boyes ED, Ward MR, Lari L, Gai PL (2013) Ann Phys Berlin 525:423–429
59. Xiong Y, Cai H, Wiley BJ, Wang J, Kim MJ, Xia Y (2007) J Am Chem Soc 129:3665–3675
60. Zhang H, Jin M, Xiong Y, Lim B, Xia Y (2012) Acc Chem Res 46:1783–1794
61. Xia X, Choi S-I, Herron JA, Lu N, Scaranto J, Peng H-C, Wang J, Mavrikakis M, Kim MJ, Xia Y (2013) J Am Chem Soc 135:15706–15709
62. Astruc D, Lu F, Aranzaes JR (2005) Angew Chem Int Ed 44:7852–7872
63. Hermans I, Spier ES, Neuenschwander U, Turrà N, Baiker A (2009) Top Catal 52:1162–1174
64. Heyns K, Paulsen H (1957) Angew Chem 69:600–608
65. Liu XY, Madix RJ, Friend CM (2008) Chem Soc Rev 37:2243–2261
66. Dimitratos N, Lopez-Sanchez JA, Hutchings GJ (2012) Chem Sci 3:20–44

67. Besson M, Gallezot P (2000) Catal Today 57:127–141
68. Kluytmans J, Markusse A, Kuster B, Marin G, Schouten J (2000) Catal Today 57:143–155
69. Mallat T, Baiker A (1994) Catal Today 19:247–283
70. Narasimharao K, Brown DR, Lee AF, Newman AD, Siril PF, Tavener SJ, Wilson K (2007) J Catal 248:226–234
71. Suwannakarn K, Lotero E, Ngaosuwan K, Goodwin JG (2009) Ind Eng Chem Res 48:2810–2818
72. Kouzu M, Nakagaito A, Hidaka J-s (2011) Appl Catal A 405:36–44
73. Zhao D, Huo Q, Feng J, Chmelka BF, Stucky GD (1998) J Am Chem Soc 120:6024–6036
74. Mbaraka IK, Shanks BH (2005) J Catal 229:365–373
75. Melero JA, Bautista LF, Morales G, Iglesias J, Briones D (2008) Energy Fuels 23:539–547
76. Chen X-R, Ju Y-H, Mou C-Y (2007) J Phys Chem C 111:18731–18737
77. Mbaraka IK, Radu DR, Lin VSY, Shanks BH (2003) J Catal 219:329–336
78. Chen D, Li Z, Wan Y, Tu X, Shi Y, Chen Z, Shen W, Yu C, Tu B, Zhao D (2006) J Mater Chem 16:1511–1519
79. Cao L, Man T, Kruk M (2009) Chem Mater 21:1144–1153
80. Dacquin JP, Lee AF, Pirez C, Wilson K (2012) Chem Commun 48:212–214
81. Martin A, Morales G, Martinez F, van Grieken R, Cao L, Kruk M (2010) J Mater Chem 20:8026–8035
82. Pirez C, Caderon J-M, Dacquin J-P, Lee AF, Wilson K (2012) ACS Catal 2:1607–1614
83. Vinu A, Gokulakrishnan N, Balasubramanian VV, Alam S, Kapoor MP, Ariga K, Mori T (2008) Chem A Eur J 14:11529–11538
84. Melero JA, Iglesias J, Morales G (2009) Green Chem 11:1285–1308
85. Carrero A, Vicente G, Rodríguez R, Linares M, del Peso GL (2011) Catal Today 167:148–153
86. Ying JY, Mehnert CP, Wong MS (1999) Angew Chem Int Ed 38:56–77
87. Lu Y (2006) Angew Chem Int Ed 45:7664–7667
88. Garg S, Soni K, Kumaran GM, Bal R, Gora-Marek K, Gupta JK, Sharma LD, Dhar GM (2009) Catal Today 141:125–129
89. Gheorghiu S, Coppens M-O (2004) AIChE J 50:812–820
90. Zhang X, Zhang F, Chan K-Y (2004) Mater Lett 58:2872–2877
91. Sun J-H, Shan Z, Maschmeyer T, Coppens M-O (2003) Langmuir 19:8395–8402
92. Dacquin J-P, Dhainaut JRM, Duprez D, Royer SB, Lee AF, Wilson K (2009) J Am Chem Soc 131:12896–12897
93. Dhainaut J, Dacquin J-P, Lee AF, Wilson K (2010) Green Chem 12:296–303
94. Wilson K, Rénson A, Clark JH (1999) Catal Lett 61:51–55
95. Rác B, Hegyes P, Forgo P, Molnár Á (2006) Appl Catal A 299:193–201
96. Yang Q, Liu J, Yang J, Kapoor MP, Inagaki S, Li C (2004) J Catal 228:265–272
97. Yang Q, Kapoor MP, Shirokura N, Ohashi M, Inagaki S, Kondo JN, Domen K (2005) J Mater Chem 15:666–673
98. Morales G, Athens G, Chmelka BF, van Grieken R, Melero JA (2008) J Catal 254:205–217

99. Margolese D, Melero JA, Christiansen SC, Chmelka BF, Stucky GD (2000) Chem Mater 12:2448–2459
100. Díaz I, Márquez-Alvarez C, Mohino F, Pérez-Pariente JN, Sastre E (2000) J Catal 193:283–294
101. Dacquin J-P, Cross HE, Brown DR, Duren T, Williams JJ, Lee AF, Wilson K (2010) Green Chem 12:1383–1391
102. Schumacher C, Gonzalez J, Wright PA, Seaton NA (2005) J Phys Chem B 110:319–333
103. Pirez C, Wilson K, Lee AF (2014) Green Chem 16:197–202
104. Ono Y, Baba T (1997) Catal Today 38:321–337
105. Hattori H (1995) Chem Rev 95:537–558
106. Albuquerque MCG, Azevedo DCS, Cavalcante CL Jr, Santamaría-González J, Mérida-Robles JM, Moreno-Tost R, Rodríguez-Castellón E, Jiménez-López A, Maireles-Torres P (2009) J Mol Catal A: Chem 300:19–24
107. Peterson GR, Scarrah WP (1984) J Am Oil Chem Soc 61:1593–1597
108. Verziu M, Coman SM, Richards R, Parvulescu VI (2011) Catal Today 167:64–70
109. Granados ML, Alonso DM, Alba-Rubio AC, Mariscal R, Ojeda M, Brettes P (2009) Energy Fuels 23:2259–2263
110. Di Serio M, Tesser R, Casale L, D'Angelo A, Trifuoggi M, Santacesaria E (2010) Top Catal 53:811–819
111. MacLeod CS, Harvey AP, Lee AF, Wilson K (2008) Chem Eng J 135:63–70
112. Montero J, Wilson K, Lee A (2010) Top Catal 53:737–745
113. Watkins RS, Lee AF, Wilson K (2004) Green Chem 6:335–340
114. Alonso DM, Mariscal R, Granados ML, Maireles-Torres P (2009) Catal Today 143:167–171
115. Verziu M, Cojocaru B, Hu J, Richards R, Ciuculescu C, Filip P, Parvulescu VI (2008) Green Chem 10:373–381
116. Zhu K, Hu J, Kübel C, Richards R (2006) Angew Chem Int Ed 45:7277–7281
117. Montero JM, Gai P, Wilson K, Lee AF (2009) Green Chem 11:265–268
118. Cantrell DG, Gillie LJ, Lee AF, Wilson K (2005) Appl Catal A 287:183–190
119. Di Serio M, Ledda M, Cozzolino M, Minutillo G, Tesser R, Santacesaria E (2006) Ind Eng Chem Res 45:3009–3014
120. Cavani F, Trifirò F, Vaccari A (1991) Catal Today 11:173–301
121. Barakos N, Pasias S, Papayannakos N (2008) Bioresour Technol 99:5037–5042
122. Fraile JM, García N, Mayoral JA, Pires E, Roldán L (2009) Appl Catal A 364:87–94
123. Cross HE, Brown DR (2010) Catal Commun 12:243–245
124. Woodford JJ, Dacquin J-P, Wilson K, Lee AF (2012) Energy Environ Sci 5:6145–6150
125. D Highley, A Bloodworth, R Bate Dolomite-mineral planning factsheet, British Geological Survey, 2006
126. Sutton D, Kelleher B, Ross JRH (2001) Fuel Process Technol 73:155–173
127. Wilson K, Hardacre C, Lee AF, Montero JM, Shellard L (2008) Green Chem 10:654–659
128. Ilgen O (2011) Fuel Process Technol 92:452–455
129. Scharff Y, Asteris D, Fédou S (2013) OCL 20:D502

130. Choi K-M, Akita T, Mizugaki T, Ebitani K, Kaneda K (2003) New J Chem 27:324–328
131. Chen J, Zhang Q, Wang Y, Wan H (2008) Adv Synth Catal 350:453–464
132. Li F, Zhang Q, Wang Y (2008) Appl Catal A 334:217–226
133. Lee AF, Wilson K (2004) Green Chem 6:37
134. Lee AF, Hackett SF, Hargreaves JS, Wilson K (2006) Green Chem 8:549–555
135. Parlett CMA, Bruce DW, Hondow NS, Lee AF, Wilson K (2011) ACS Catal 1:636–640
136. Parlett CMA, Bruce DW, Hondow NS, Newton MA, Lee AF, Wilson K (2013) Chem Cat Chem 5:939–950
137. Hackett SF, Brydson RM, Gass MH, Harvey I, Newman AD, Wilson K, Lee AF (2007) Angew Chem 119:8747–8750
138. Baeza J, Calvo L, Gilarranz M, Mohedano A, Casas J, Rodriguez J (2012) J Catal 293:85–93
139. Wang X, Yang H, Feng B, Hou Z, Hu Y, Qiao Y, Li H, Zhao X (2009) Catal Lett 132:34–40
140. Yudha S, Dhital RN, Sakurai H (2011) Tetrahedron Lett 52:2633–2637
141. Dun R, Wang X, Tan M, Huang Z, Huang X, Ding W, Lu X (2013) ACS Catal 3:3063–3066
142. Weldon MK, Friend CM (1996) Chem Rev 96:1391–1412
143. Lee AF, Chang Z, Ellis P, Hackett SF, Wilson K (2007) J Phys Chem C 111:18844–18847
144. Naughton J, Pratt A, Woffinden CW, Eames C, Tear SP, Thompson SM, Lee AF, Wilson K (2011) J Phys Chem C 115:25290–25297
145. Davis JL, Barteau MA (1987) Surf Sci 187:387–406
146. Davis JL, Barteau MA (1990) Surf Sci 235:235–248
147. Zaera F (2003) Catal Lett 91:1–10
148. Keresszegi C, Burgi T, Mallat T, Baiker A (2002) J Catal 211:244–251
149. Lee AF (2001) Abstr Pap Am Chem Soc 221:U335–U336
150. Grunwaldt J-D, Keresszegi C, Mallat T, Baiker A (2003) J Catal 213:291–295
151. Keresszegi C, Grunwaldt J-D, Mallat T, Baiker A (2004) J Catal 222:268–280
152. Grunwaldt J-D, Caravati M, Baiker A (2006) J Phys Chem B 110:25586–25589
153. Mondelli C, Ferri D, Grunwaldt J-D, Krumeich F, Mangold S, Psaro R, Baiker A (2007) J Catal 252:77–87
154. Keresszegi C, Ferri D, Mallat T, Baiker A (2005) J Catal 234:64–75
155. Caravati M, Meier DM, Grunwaldt J-D, Baiker A (2006) J Catal 240:126–136
156. Mallat T, Baiker A (1995) Catal Today 24:143–150
157. Mallat T, Bodnar Z, Hug P, Baiker A (1995) J Catal 153:131–143
158. Parlett CMA, Keshwalla P, Wainwright SG, Bruce DW, Hondow NS, Wilson K, Lee AF (2013) ACS Catal 3:2122–2129
159. Parlett CMA, Durndell LJ, Wilson K, Bruce DW, Hondow NS, Lee AF (2014) Catal Commun 44:40–45
160. Naughton J, Lee AF, Thompson S, Vinod CP, Wilson K (2010) Phys Chem Chem Phys 12:2670–2678
161. Ping EW, Pierson J, Wallace R, Miller JT, Fuller TF, Jones CW (2011) Appl Catal A 396:85–90

162. Newton MA, Jyoti B, Dent AJ, Fiddy SG, Evans J (2004) Chem Commun 2004:2382–2383
163. Ferri D, Baiker A (2009) Top Catal 52:1323–1333
164. Newton MA (2009) Top Catal 52:1410–1424
165. Newton MA, Dent AJ, Fiddy SG, Jyoti B, Evans J (2007) Catal Today 126:64–72
166. Lee AF, Ellis CV, Naughton JN, Newton MA, Parlett CM, Wilson K (2011) J Am Chem Soc 133:5724–5727
167. Parlett CMA, Gaskell CV, Naughton JN, Newton MA, Wilson K, Lee AF (2013) Catal Today 205:76–85
168. Gaskell CV, Parlett CMA, Newton MA, Wilson K, Lee AF (2012) ACS Catal 2:2242–2246
169. Lee AF, Naughton JN, Liu Z, Wilson K (2012) ACS Catal 2:2235–2241
170. Tauster S (1987) Acc Chem Res 20:389–394
171. Ghedini E, Menegazzo F, Signoretto M, Manzoli M, Pinna F, Strukul G (2010) J Catal 273:266–273
172. Hicks RF, Qi H, Young ML, Lee RG (1990) J Catal 122:295–306
173. Beckers J, Rothenberg G (2010) Green Chem 12:939–948
174. Inumaru K, Nakamura K, Ooyachi K, Mizutani K, Akihara S, Yamanaka S (2010) J Porous Mater 18:455–463
175. Liu Y-M, Cao Y, Yi N, Feng W-L, Dai W-L, Yan S-R, He H-Y, Fan K-N (2004) J Catal 224:417–428
176. Kim T-W, Kleitz F, Paul B, Ryoo R (2005) J Am Chem Soc 127:7601–7610
177. Zhao D, Feng J, Huo Q, Melosh N, Fredrickson GH, Chmelka BF, Stucky GD (1998) Science 279:548–552
178. Sneh O, George SM (1995) J Phys Chem 99:4639–4647
179. Kim JM, Kwak JH, Jun S, Ryoo R (1995) J Phys Chem 99:16742–16747
180. Kaneda K, Fujii M, Morioka K (1996) J Org Chem 61:4502–4503
181. Polshettiwar V, Varma RS (2009) Org Biomol Chem 7:37–40
182. Peterson KP, Larock RC (1998) J Org Chem 63:3185–3189
183. Behafarid F, Ono LK, Mostafa S, Croy JR, Shafai G, Hong S, Rahman TS, Bare SR, Cuenya BR (2012) Phys Chem Chem Phys 14:11766–11779
184. Yuan Q, Duan HH, Li LL, Sun LD, Zhang YW, Yan CH (2009) J Colloid Interface Sci 335:151–167
185. Lykhach Y, Staudt T, Lorenz MPA, Streber R, Bayer A, Steinrück H-P, Libuda J (2010) Chem Phys Chem 11:1496–1504
186. Tana, Zhang M, Li J, Li H, Li Y, Shen W (2009) Catal Today 148:179–183
187. Bensalem A, Bozon-Verduraz F, Perrichon V (1995) J Chem Soc Faraday Trans 91:2185–2189
188. Harrison B, Diwell A, Hallett C (1988) Plat Met Rev 32:73–83
189. Zhu Y, Zhang S, Shan J-j, Nguyen L, Zhan S, Gu X, Tao F (2013) ACS Catal 3:2627–2639
190. Maillet T, Madier Y, Taha R, Barbier J Jr, Duprez D (1997) Stud Surf Sci Catal 112:267–275
191. Oh S-H, Hoflund GB (2006) J Phys Chem A 110:7609–7613
192. Badri A, Binet C, Lavalley J-C (1996) J Chem Soc Faraday Trans 92:1603–1608
193. Lee AF, Gee JJ, Theyers HJ (2000) Green Chem 2:279–282

194. Wenkin M, Ruiz P, Delmon B, Devillers M (2002) J Mol Catal A: Chem 180:141–159
195. Alardin F, Delmon B, Ruiz P, Devillers M (2000) Catal Today 61:255–262
196. Keresszegi C (2004) J Catal 225:138–146
197. Pinxt H, Kuster B, Marin G (2000) Appl Catal A 191:45–54
198. Dimitratos N, Porta F, Prati L, Villa A (2005) Catal Lett 99:181–185
199. Prati L, Villa A, Campione C, Spontoni P (2007) Top Catal 44:319–324
200. Enache DI, Edwards JK, Landon P, Solsona-Espriu B, Carley AF, Herzing AA, Watanabe M, Kiely CJ, Knight DW, Hutchings GJ (2006) Science 311:362–365
201. Long J, Liu H, Wu S, Liao S, Li Y (2013) ACS Catal 3:647–654
202. Griffin MB, Rodriguez AA, Montemore MM, Monnier JR, Williams CT, Medlin JW (2013) J Catal 307:111–120
203. Liu Z, Xu J, Zhang H, Zhao Y, Yu B, Chen S, Li Y, Hao L, Green Chem 2013
204. Wang R, Wu Z, Chen C, Qin Z, Zhu H, Wang G, Wang H, Wu C, Dong W, Fan W (2013) Chem Commun 49:8250–8252
205. Hou W, Dehm NA, Scott RW (2008) J Catal 253:22–27
206. Mertens P, Corthals S, Ye X, Poelman H, Jacobs P, Sels B, Vankelecom I, De Vos D (2009) J Mol Catal A: Chem 313:14–21
207. Hutchings GJ (2005) Catal Today 100:55–61
208. Lee AF, Ellis CV, Wilson K, Hondow NS (2010) Catal Today 157:243–249
209. Scott RW, Wilson OM, Oh S-K, Kenik EA, Crooks RM (2004) J Am Chem Soc 126:15583–15591
210. Scott RW, Sivadinarayana C, Wilson OM, Yan Z, Goodman DW, Crooks RM (2005) J Am Chem Soc 127:1380–1381
211. Dash P, Bond T, Fowler C, Hou W, Coombs N, Scott RWJ (2009) J Phys Chem C 113:12719–12730
212. Maclennan A, Banerjee A, Hu Y, Miller JT, Scott RWJ (2013) ACS Catal 3:1411–1419

CHAPTER 2

Catalytic Hydroprocessing of Liquid Biomass for Biofuels Production

STELLA BEZERGIANNI

2.1 INTRODUCTION

The depletion of world petroleum reserves and the increased concern on climate change has stimulated the recent interest in biofuels. The most common biofuels are based on energy crops and their products, i.e. vegetable oil for Fatty Acid Methyl Esters (FAME) biodiesel [1] and sugars/starch for bioethanol. However these first generation biofuels and associated production technologies face several considerations related to their economic and social implications regarding energy crops cultivation, by-products disposal, necessity for large investments to ensure competitiveness and the "food versus fuel" debate.

As a result, second generation biofuel technologies have been developed to overcome the limitations of first generation biofuels production [2]. The goal of second generation biofuel processes is to extend biofuel production capacity by incorporating residual biomass while increasing

Stella Bezergianni (2013). Catalytic Hydroprocessing of Liquid Biomass for Biofuels Production, Liquid, Gaseous and Solid Biofuels - Conversion Techniques, Prof. Zhen Fang (Ed.), ISBN: 978-953-51-1050-7, InTech, DOI: 10.5772/52649. Licensed under Creative Commons Attribution 3.0 Unported License, http://creativecommons.org/licenses/by/3.0/.

sustainability. This residual biomass consists of the non-food parts of food crops (such as stems, leaves and husks) as well as other non-food crops (such as switch grass, jatropha, miscanthus and cereals that bear little grain). Furthermore the residual biomass potential is further augmented by industrial and municipal organic waste such as skins and pulp from fruit pressing, waste cooking oil etc. One such technology is catalytic hydro-processing, which is an alternative conversion technology of liquid bio-mass to biofuels that is lately raising a lot of interest in both the academic and industrial world and is the proposed subject of this chapter.

Catalytic hydroprocessing is a key process in petrochemical industry for over a century enabling heteroatom (sulfur, nitrogen, oxygen, metals) removal, saturation of olefins and aromatics, as well as isomerization and cracking [3]. Due to the numerous applications of catalytic hydroprocess-ing, there are several catalytic hydroprocessing units in a typical refinery including distillate hydrotreaters and hydrocrackers (see Figure 1). As a result several refinery streams are treated with hydrogen in order to im-prove final product quality including straight-run naphtha, diesel, gas-oils etc. The catalytic hydroprocessing technology is evolving through the new catalytic materials that are being developed. Even though hydroprocess-ing catalysts development is well established [4], the growing demand of petroleum products and their specifications, which are continuously be-coming stricter, have created new horizons in the catalyst development in order to convert heavier and lower quality feedstocks [5]. Furthermore the expansion of the technology to bio-based feedstocks has also broadened the R&D spam of catalytic hydrotreatment.

Catalytic hydroprocessing of liquid biomass is a technology that offers great flexibility to the continuously increasing demands of the biofuels market, as it can convert a wide variety of liquid biomass including raw vegetable oils, waste cooking oils, animal fats as well as algal oils into biofuels with high conversion yields. In general this catalytic process tech-nology allows the conversion of triglycerides and lipids into paraffins and iso-paraffins within the naphtha, kerosene and diesel ranges. The products of this technology have improved characteristics as compared to both their fossil counterparts and the conventional biofuels including high heating value and cetane number, increased oxidation stability, negligible acid-ity and increased saturation level. Besides the application of this catalytic technology for the production of high quality paraffinic fuels, catalytic hydroprocessing is also an effective technology for upgrading intermediate products of solid biomass conversion technologies such as pyrolysis oils and Fischer-Tropsch wax (Figure 2). The growing interest and investments of the petrochemical, automotive and aviation industries to the biomass catalytic hydroprocessing technology shows that this technology will play an important role in the biofuels field in the immediate future.

In the sections that follow, the basic technical characteristics of catalytic hydrotreatment are presented including a description of the process, reactions, operating parameters and feedstock characteristics. Furthermore key applications of catalytic hydroprocessing of liquid biomass are outlined based on different feedstocks including raw vegetable oils, waste cooking oils, pyrolysis oils, Fischer-Tropsch wax and algal oil, and some successful demonstration activities are also presented.

2.2 TECHNICAL CHARACTERISTICS OF CATALYTIC HYDROTREATMENT

The catalytic hydrotreatment of liquid biomass converts the contained triglycerides/lipids into hydrocarbons at high temperatures and pressures over catalytic material under excess hydrogen atmosphere. The catalytic hydrotreatment of liquid biomass process is quite similar to the typical process applied to petroleum streams, as shown in Figure 3. A typical catalytic hydrotreatment unit consists of four basic sections: a) feed preparation, b) reaction, c) product separation and d) fractionation.

In the feed preparation section the liquid biomass feedstock is mixed with the high pressure hydrogen (mainly from gas recycle with some additional fresh make-up hydrogen) and is preheated before it enters the reactor section. The reactor section consists normally of two hydrotreating reactors, a first guard mild hydrotreating reactor and a second one where the main hydrotreating reactions take place. Each reactor contains two or more catalytic beds in order to maintain constant temperature profile throughout the reactor length. Within the reactor section all associated reactions take place, which will be presented in more detail at a later paragraph.

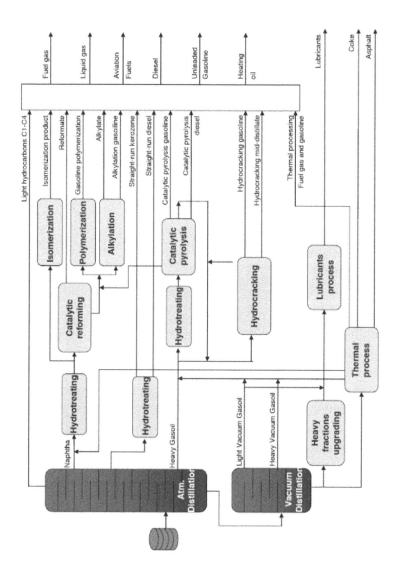

FIGURE 1: Catalytic hydroprocessing units within a refinery, including distillate hydrotreating and hydrocracking

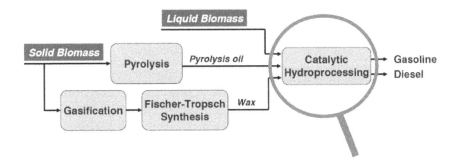

FIGURE 2: Catalytic hydroprocessing for biomass conversion and upgrading towards fuels production

The reactor product then enters the separator section where, after it is cooled down, it enters the high pressure separator (HPS) flash drum in which the largest portion of the gas and liquid product molecules are separated. The gas product of the HPS includes the excess hydrogen that has not reacted within the reactor section as well as the side products of the reactions including CO, CO_2, H_2S, NH_3 and H_2O. The liquid product of the HPS is lead to a second flash drum, the low pressure separator (LPS), for removing any residual gas contained in the liquid product, and subsequently is fed to a fractionator section. The fractionator section provides the final product separation into the different boiling point fractions that yield the desired products including off-gas, naphtha, kerosene and diesel. The heaviest molecules return from the bottom of the fractionator into the reactor section as a liquid recycle stream.

In order to improve the overall efficiency, a liquid recycle stream is also incorporated, which in essence consists of the heavy molecules that were not converted. The gas product from the HPS and LPS, after being treated to remove the excess NH_3, H_2S, CO and CO_2, is compressed and fed back to the reactor section as a gas recycle stream in order to maintain a high pressure hydrogen atmosphere within the reactor section.

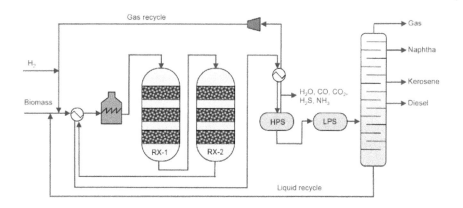

FIGURE 3: A typical process diagram of catalytic hydrotreatment of liquid biomass

2.2.1 REACTION MECHANISMS

Several types of reactions take place during catalytic hydrotreatment of liquid biomass, based on the type of biomass processed, operating conditions and catalyst employed. The types of reactions that liquid biomass undergoes during catalytic hydroprocessing include: a) cracking, b) saturation, c) heteroatom removal and d) isomerization, which are described in more detail in the following section.

2.2.1.1 CRACKING

As the molecules included in the various types of liquid biomass can be relatively large and complicated, cracking reactions are desired to convert them into molecules of the size and boiling point range of conventional fuels, mainly gasoline, kerosene and diesel. A characteristic reaction that occurs during catalytic hydrotreating of oils / fats is the cracking of triglycerides into its consisting fatty acids (carboxylic acids) and propane as shown in Scheme 1 [5][6]. This reaction is critical as it converts the initial large triglycerides molecules of boiling point over 600°C into mid-distillate range molecules (naphtha, kerosene and diesel).

Other cracking reactions may take place however such as those described in Schemes 2 and 3, depending on the type of molecules present in the feedstock. For example Scheme 2 is a cracking reaction which may occur during catalytic hydrotreatment of pyrolysis oil which includes polyaromatic and aromatic compounds. Alternatively Scheme 3 may follow deoxygenation of carboxylic acids on the produced long chain paraffinic molecules, leading to smaller chain paraffins, during the upgrading of Fischer-Tropsch wax.

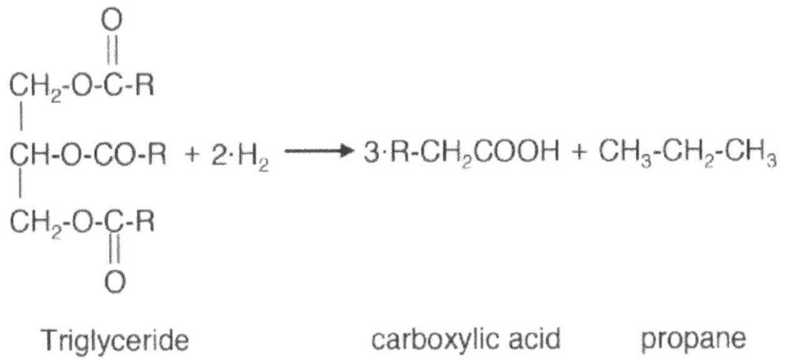

$$
\begin{array}{c}
\quad\quad O \\
\quad\quad \| \\
CH_2\text{-}O\text{-}C\text{-}R \\
| \\
CH\text{-}O\text{-}CO\text{-}R \;+\; 2 \cdot H_2 \longrightarrow 3 \cdot R\text{-}CH_2COOH + CH_3\text{-}CH_2\text{-}CH_3 \\
| \\
CH_2\text{-}O\text{-}C\text{-}R \\
\quad\quad\quad \| \\
\quad\quad\quad O
\end{array}
$$

Triglyceride   carboxylic acid   propane

SCHEME 1

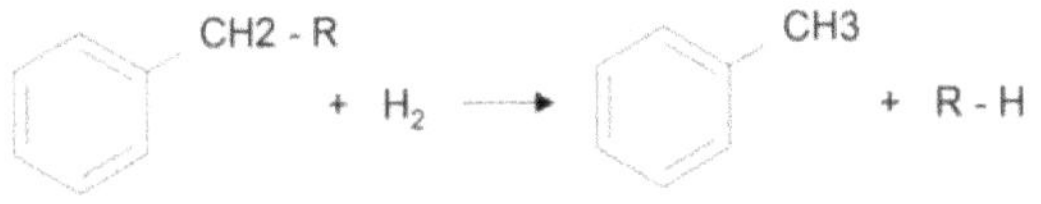

SCHEME 2

$$R\text{-}R' + H_2 \longrightarrow R\text{-}H + R'\text{-}H$$

SCHEME 3.

2.2.1.2 SATURATION

Saturation reactions are strongly associated with catalytic hydrotreating as the introduction of excess hydrogen allows the breakage of double C-C bonds and their conversion to single bonds, as shown in the following reactions. In particular the saturation of unsaturated carboxylic acids into saturated ones depicted in Scheme 4, is a key reaction occurring in lipid feedstocks. Furthermore other saturation reactions lead to the formation of naphthenes by converting unsaturated cyclic compounds and aromatic compounds as in Scheme 5 and 6, which are likely to occur during upgrading of pyrolysis oils.

$$RCH = CH\text{-}COOH + H_2 \longrightarrow RCH_2CH_2COOH$$

SCHEME 4

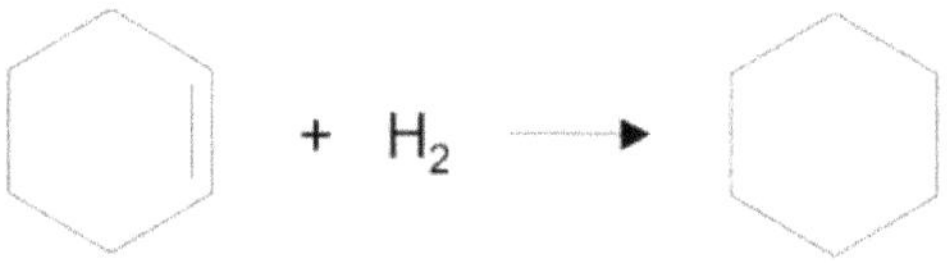

SCHEME 5

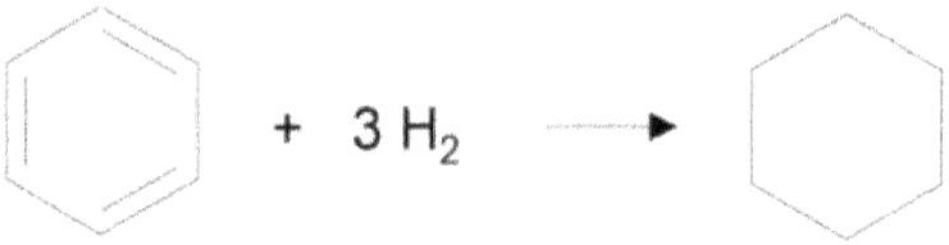

SCHEME 6

As a result of this reaction the produced saturated molecules are less active and less prone to polymerization and oxidation reactions, mitigating the sediment formation and corrosion phenomena appearing in engines.

1.2.1.3 HETEROATOM REMOVAL

Heteroatoms are atoms other than carbon (C) and hydrogen (H) and are often encountered into bio- and fossil- based feedstocks. They include sulfur (S), nitrogen (N) and in the case of bio-based feedstocks oxygen (O). In particular oxygen removal is of outmost importance as the presence of oxygen reduces oxidation stability (due to carboxylic and carbonylic double bonds), increases acidity and corrosivity (due to the presence of water) and even reduces the heating value of the final biofuels. The main deoxygenation reactions that take place include deoxygenation, decarbonylation and decarboxylation presented in Schemes 7, 8 and 9 respectively [7]. The main products of deoxygenation reactions include n-paraffins, while H_2O, CO_2 and CO are also produced, but can be removed with the excess hydrogen within the flash drums of the product separation section. It should be noted however that these particular reactions give the paraffinic nature of the produced biofuels, and for this reason the hydrotreated products are often referred to as paraffinic fuels (e.g. paraffinic jet, paraffinic diesel etc)

$$R\text{-}CH_2COOH + 3 \cdot H_2 \longrightarrow R\text{-}CH_2CH_3 + 2 \cdot H_2O$$

SCHEME 7

$$R\text{-}CH_2COOH + H_2 \longrightarrow R\text{-}CH_3 + CO + H_2O$$

SCHEME 8

$$R\text{-}CH_2COOH + H_2 \longrightarrow R\text{-}CH_3 + CO_2$$

SCHEME 9

The other heteroatoms, i.e. S and N are removed according to the classic heteroatom removal mechanisms of the fossil fuels in the form of gaseous H_2S and NH_3 respectively.

1.2.1.4 ISOMERIZATION

The straight chain paraffinic molecules resulting from the aforementioned reactions, even though they offer increased cetane number, heating value and oxidation stability in the biofuels which contain them, they also degrade their cold flow properties. In order to improve the cold flow properties, isomerization reactions are also required, which normally take place during a second step/reactor as they require a different catalyst. Some examples of isomerization reactions are given in Schemes 10 and 11.

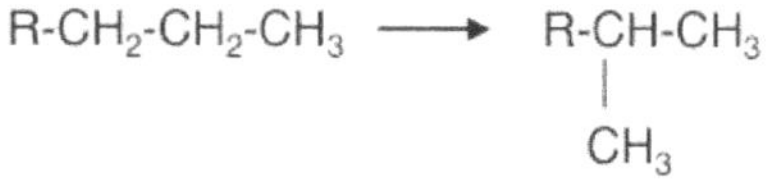

$$R\text{-}CH_2\text{-}CH_2\text{-}CH_3 \longrightarrow R\text{-}\underset{\underset{\displaystyle CH_3}{|}}{C}H\text{-}CH_3$$

SCHEME 10

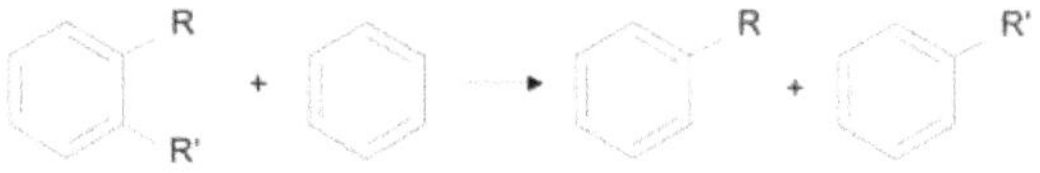

SCHEME 11

2.2.2 HYDROPROCESSING CATALYSTS

Catalytic hydroprocessing of liquid biomass is a technology currently under developed and there is a lot of room for optimization. For example there are not many commercial catalysts specifically designed and developed for such applications, while conventional commercial catalysts, employed for catalytic hydroprocessing of refinery streams, are used instead. Common hydrotreating catalysts employed contain active metals on alumina substrate with increased surface area. The most known commercial catalysts employ Cobalt and Molybdenum (CoMo) or Nikel and Molybdenum (NiMo) in alumina substrate (Al_2O_3) as shown in Figure 4.

Hydrotreating catalysts are dual action catalytic material, triggering both hydrogenation and cracking/isomerization reactions. On one hand hydrogenation takes place on the active metals (Mo, Ni, Co, Pd, Pt) which catalyze the feedstock molecules rendering them more active when subject to cracking and heteroatom removal, while limiting coke formation on the catalyst. Furthermore hydrogenation supports cracking by forming an active olefinic intermediate molecule via dehydrogenation. On the other hand both cracking and isomerization reactions take place in acidic environment such as amorphous oxides ($SiO_2 - Al_2O_3$) or crystalline zeolites (mainly z-zeolites) or mixtures of zeolites with amorphous oxides.

During the first contact of the feedstock molecules with the catalyst, a temperature increase is likely to develop due to the exothermic reactions that occur. However, during the continuous utilization of the catalyst and coke deposition, the catalyst activity eventually reduces from 1/3 to 1/2 of its initial one. The catalyst deactivation rate mainly depends on temperature and hydrogen partial pressure. Increased temperatures accelerate catalyst deactivation while high hydrogen partial pressure tends to mitigate catalyst deactivation rate. Most of the catalyst activity can be recovered by catalyst regeneration.

The selection of a suitable hydroprocessing catalyst is a critical step defining the hydroprocessing product yield and quality as well as the operating cycle time of the process in petroleum industry [5]. However the hydrotreating catalyst selection for biomass applications is particularly crucial and challenging for two reasons: a) catalyst activity varies significantly, as commercial catalysts are designed for different feedstocks, i.e. feedstocks with high sulfur concentration, heavy feedstocks (containing large molecules), feedstocks with high oxygen concentration etc, and b) there are currently no commercial hydroprocessing catalysts available for lipid feedstocks and other intermediate products of biomass conversion processes (e.g. pyrolysis biooil), and thus commercial hydrotreating catalysts need to be explored and evaluated as different catalyst have different yields (Figure 5) and different degradation rate [8].

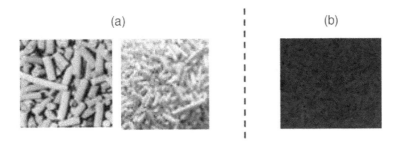

FIGURE 4: Typical hydrotreating catalysts (a) before use and (b) after use

Nevertheless, significant efforts have been directed t0wards developing special hydrotreating catalysts for converting/upgrading liquid biomass to biofuels [9-12].

2.2.3 OPERATING PARAMETERS

As it was mentioned earlier, the choice of catalyst and operating parameters affect the reactions that take place within the hydroprocessing reactor. The key operating parameters of hydroprocessing include the reactor temperature, hydrogen partial pressure, liquid hourly space velocity and hydrogen feed-rate.

2.2.3.1 TEMPERATURE

Most catalytic hydrotreating and hydrocracking reactors operate between 290-450°C. The temperature range is selected according the type of catalyst and feedstock type to be processed. In the first stages of the catalyst life (after its loading in the reactor) the temperature is normally kept low as the catalyst activity is already very high. However as time progresses and the catalyst deactivates and cokes, the temperature is gradually increased to overcome the loss of catalyst activity and to maintain the desired product yield and quality.

2.2.3.2 HYDROGEN PARTIAL PRESSURE

Hydrogen partial pressure affects significantly the hydrotreating reactions as well as the catalyst deactivation. The catalyst deactivation rate is inverse proportional to the hydrogen partial pressure and to hydrogen feed-rate. However high hydrogen partial pressures correspond to high operational costs, which rise even higher for high olefinic feedstocks that exhibit higher hydrogen consumption due to the saturation reactions. Therefore hydrogen partial pressure should be balanced with the catalyst activity and catalyst life expectancy in order to optimize the overall process.

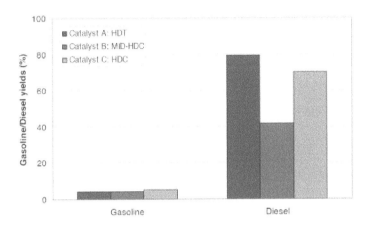

FIGURE 5: Catalyst comparison based on gasoline and diesel yields for WCO hydrotreating [8]. (Reprinted from Fuel, 93, S. Bezergianni, A. Kalogianni, A. Dimitriadis, Catalyst evaluation for waste cooking oil hydroprocessing, 638-641, 2012, with permission from Elsevier).

2.2.3.3 LIQUID HOURLY SPACE VELOCITY

Liquid hourly space velocity (LHSV) is defined as the ratio of the liquid mass feed-rate (gr/h) over the catalyst mass (gr) and as a result is expressed in hr^{-1}. In fact the inverse of LHSV is proportional to the residence time of the liquid feed in the reactor. In essence the higher the liquid hourly space velocity, the less time is available for the contact of the feed molecules of the reaction mixture with the catalyst, thus the less the conversion. However, maintaining large LHSV imposes a faster degradation of the catalyst therefore in industrial applications the LHSV is maintained in as high values as it is practically possible.

2.2.3.4 HYDROGEN FEED-RATE

The hydrogen feed-rate is another important parameter as it also defines hydrogen partial pressure depending on the hydrogen consumption of each application. It actually favours both heteroatom removal and saturation reaction rates. However, as hydrogen cost defines the overall unit operating cost, hydrogen feed-rate is normally optimized depending on the system requirements. Furthermore the use of renewable energy sources for hydrogen production is also envisioned as a potential cost improvement option.

2.3 FEED SAND PRODUCTS

Even though liquid biomass is currently being exploited as a renewable feedstock for fuels production, its characteristics are far beyond suitable for its use as fuel. More specifically liquid biomass, just as other types of biomass, has a small H/C ratio and high oxygen content, lowering its heating value and increasing CO and CO_2 emissions during its combustion. Moreover liquid biomass contains water, which can cause corrosion in the downstream processing units if it's not completely removed, or even in the engine parts where its final products are utilized. In addition to the above, liquid biomass has an increased concentration in oxygenated compounds, mainly acids, aldehydes, ketones etc, which not only reduce the heating value, but also decrease the oxidation stability and increase the acidity of the produced biofuels. For all the aforementioned reasons it is imperative that liquid biomass should be upgraded and specifically that its H/C should be increased while the water and oxygen removed.

The effectiveness of catalytic hydroprocessing towards improving these problematic characteristics of liquid biomass is presented in Table 1, where the H/C ratio, the oxygen content and density before and after catalytic hydrotreatment of basic liquid biomass types are given. The H/C ratio exhibits a significant increase that exceeds 50% in all cases. This is due to the substitution of the heteroatoms by hydrogen atoms as well as in the saturation of double bonds that enriches the H/C analogy. The oxygen content (including the oxygen contained in the water) from over 15%wt can be decreased down to 5wppm. Actually the deep deoxygenation achieved via catalytic hydrotreatment is the most significant contribution of this biomass conversion technology, as it improves significantly the oxidation stability of the final biofuels. Furthermore significant improvement is also observed in the biomass density, which is never below 0.9 kg/l while after hydrotreatment it reduces to values less than 0.8 kg/l

TABLE 1: Effect of catalytic hydrotreatment on the liquid biomass characteristics

	Liquid biomass (unprocessed)	Hydrotreated liquid biomass and produced biofuels
H/C ratio	0.08 – 0.1	0.13 – 0.18
Oxygen content (%wt)	15 - 40	10^{-4} – 3
Density (kg/l)	0.9 – 1.05	0.75 – 0.8

Catalytic hydroprocessing has been proven as the most efficient technology for the upgrading of liquid biomass as it achieves to increase the H/C ratio and to remove oxygen and water. However the effectiveness of this technology is also shown in other parameters. For example the distillation curve of raw liquid biomass shows that over 90% of its molecules have boiling points exceeding 600°C and only 5% are within diesel range (220-360°C), while after catalytic hydrotreatment upgrading most of 90% of the product molecules are within diesel range [13].

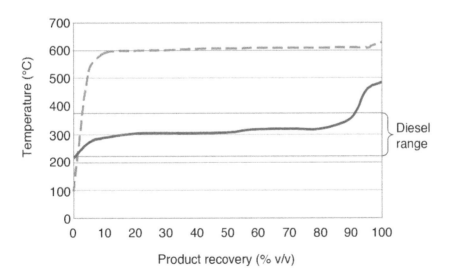

FIGURE 6: Distillation curves of untreated WCO (dashed) and catalytically hydrotreated WCO (solid)

TABLE 2: Fatty acid composition of most common vegetable oils [14][15]

		C8:0	C10:0	C12:0	C14:0	C16:0	C16:1	C18:0	C18:1	C18:2	C18:3	C20:0/C22:0	C20:1/C22:1
EU	Rapeseed oil	0.0	0.0	0.0	0.0	3.5	1.0	1.5	12.5	15.0	7.5	9.0	50.0
	Soybean oil	0.0	0.0	0.0	0.3	8.2	0.5	4.5	25.0	49.0	5.0	7.5	0.0
	Sunflower oil	0.0	0.0	0.0	0.0	6.0	0.0	4.2	18.8	69.3	0.3	1.4	0.0
	Corn oil	0.0	0.0	0.0	1.0	9.0	1.5	2.5	40.0	45.0	0.0	0.0	1.0
non-EU	Palm oil	0.0	0.0	0.0	3.5	39.5	0.0	3.5	47.0	6.5	0.0	0.0	0.0
	Peanut oil	0.0	0.0	0.0	0.5	8.0	1.5	3.5	51.5	27.5	0.0	7.5	0.0
	Canola oil	0.0	0.0	0.1	0.1	4.7	0.1	1.6	65.9	21.2	5.2	1.2	0.0
	Castor oil	0.0	0.1	0.2	10.6	1.4	9.5	29.7	29.7	41.3	3.3	3.8	0.0

In the following sections the basic types of liquid biomass and their corresponding products via catalytic hydrotreatment are presented.

2.3.1 RAW VEGETABLE OILS CONVERSION TO PARAFFINIC BIOFUELS

Vegetable oils are the main feedstock for the production of first generation biofuels, which can offer several CO_2 benefits and limit the consumption of fossil fuels. Raw vegetable oils consist of fatty acid triglycerides, the consistency of which depends on their origin (i.e. plant type) as shown in Table 2. Their production, however, is competing for the cultivated areas that were originally dedicated for the production of food and feed crops. As a result the production and utilization of vegetable oils for biofuels production has instigated the "food vs. fuel" debate. For this reason traditional energy crops (soy, cotton, etc) with low oil yield per hectare are being substituted by new energy crops (eg. jatropha, palm, castor etc).

Catalytic hydrotreatment was explored for conversion of vegetable oils in the early 90's. The investigation of the hydrogenolysis of various vegetable oils such as maracuja, buritimtucha, and babassu oils over a Ni–Mo/γ-Al$_2$O$_3$ catalyst as well as the effect of temperature and pressure on its effectiveness was firstly investigated [16][17]. The reaction products included a gas product rich in the excess hydrogen, carbon monoxide, carbon dioxide and light hydrocarbons as well as a liquid organic product of paraffinic nature. In more detail these studies showed the conversion of triglycerides into carboxyl oxides and then to high qualityhydrocarbons via decarboxylation and decarbonylation reactions. Rapeseed oil hydroprocessing was also studied in lab-scale reactor for temperatures 310° and 360°C and hydrogen pressures of 7 and 15 MPa using three different Ni–Mo/alumina catalysts [18]. These products contained mostly n-heptadecane and n-octadecane accompanied by low concentrations of other n-alkanes and i-alkanes [19].

2.3.2 WASTE COOKING OILS CONVERSION TO PARAFFINIC BIOFUELS

Even though vegetable oils are the main feedstock for the production of first generation biofuels, soon their production has troubled the public opinion due to their abated sustainability and to their association with the food vs. fuel debate. As a result the technology hasshifted towards the exploitation of both solid and liquid residual biomass. Waste Cooking Oils (WCOs) is a type of residual biomass resulting from frying with typical vegetable fryingoils (e.g. soybean-oil, corn-oil, olive-oil, sesame-oil etc). WCOs have particular problems regarding their disposal. In particular grease may result in coating of pipelines within the residential sewage system and is one of the most common causes of clogs and sewage spills. Furthermore, in the cases that sewage leaks into the environment, WCOs can cause human and environmental health problems because of the pathogens contained. It has been estimated that by disposing 1 lit of WCO, over 1,000,000 of liters of water can be contaminated,which is estimated as the average demand of a single person for 14 years.

Catalytic hydroprocessing of WCO was studied as an alternative approach of producing 2nd generation biofuels [20-24]. Initially catalytic hydrocracking was investigated over commercial hydrocracking catalysts leading not only to biodiesel but also to lighter products such as biogasoline [20], employing a continuous-flow catalytic hydroprocessing pilot-plant with afixed-bed reactor. During this study several parameters were considered including hydrocracking temperature (350-390°C) and liquid hourly space velocity or LHSV (0.5-2.5 hr-1) under high pressure (140 bar), revealing that the conversion is favoured by high reaction temperature and low LHSV. Lower and medium temperatures, however, were more suitable for biodiesel production while higher temperatures offered better selectivity for biogasolineproduction. Furthermore, heteroatom removal (S, N and particularly O) was increased while saturation of double bonds was decreased with increasing hydrocracking temperature, indicating the necessity of a pre-treatment step.

However catalytic hydrotreatment was later examined in more detail as a more promising technology particularly for paraffinic biodiesel production (Figure 1). The same team has studied the effect of temperature (330-

398°C) on the product yields and heteroatom removal[21]. The study was conducted in the same pilot plant utilizing a commercial $NiMo/Al_2O_3$ hydrotreating catalyst over lower pressure (80 bar). According to this study, the hydrotreatingtemperature is the key operating parameter which defines the catalyst effectiveness and life. In fact lower temperatures (330°C) favour diesel production and selectivity. Sulfur and nitrogen removal were equally effective at all temperatures, while oxygen removal and saturationof double bonds were favoured by hydrotreating temperature. The same team also studied the effect of the other three operating parameters i.e. pressure, LHSV and H_2/WCO ratio [22]. Moreover they also studied the hydrocarbon content of the products [23] qualitatively via two-dimensional chromatography and quantitatively via Gas Chromatographywith Flame Ionization Detector (GC-FID), which indicated the presence of C_{15}-C_{18} paraffins. Interestingly this study showed that as hydrotreating temperature increases, the contentof normal paraffins decreases while of iso-paraffins increases, revealing that isomerization reactions are favoured by temperature.

FIGURE 7: Catalytic hydrotreatment of WCO to 2nd generation biodiesel

The total liquid product of WCO catalytic hydrotreatment was further investigated in terms of its percentage that contains paraffins within the diesel boiling point range (220-360°C)[24]. The properties of WCO, hydrotreated WCO (total liquid product) and the diesel fractionof the hydrotreated WCO are presented in Table 3. Based on this study the overall yieldof the WCO catalytic hydrotreatment technology was estimated over 92%v/v. The properties of the new 2nd generation paraffinic diesel product indicated a high-quality diesel with highheating value (49MJ/kg) and high cetane index (77) which is double of the one of fossil diesel.An additional advantage of the new biodiesel is its oxidation stability (exceeding 22hrs) and negligible acidity, rendering it as a safe biofuel, suitable for use in all engines. The properties and potential of the new biodiesel were further studied [25], for evaluating different fractions of the total liquid product and their suitability as an alternative diesel fuel.

TABLE 3: Basic properties of waste cooking oil, hydrotreated waste cooking oil and final biodiesel

		WCO Hydrotreated	WCO	Final biodiesel
Density	gr/cm^3	0.896	0.7562	0.7869
C	wt%	76.74	84.59	86.67
H	wt%	11.61	15.02	14.74
S	wppm	38	11.80	1.54
N	wppm	47.42	0.77	1.37
O	wt%	14.57	0.38	0
Recovery 0%	°C	431.6	195.6	234.1
Recovery 10%	°C	556.4	287.4	294.1
Recovery 30%	°C	599	304.0	296.8
Recovery 50%	°C	603.2	314.4	298.3
Recovery 70%	°C	609	319.0	300
Recovery 90%	°C	612.4	320.4	298.3
Recovery 100%	°C	727.2	475.4	306.2

TABLE 4: Properties of different pyrolysis oils according to literature

Properties	Test Methods	Types of Pyrolysis Biooils						
		[26]	[27]	[28]	[29]	[30]	[31]	[32]
pH	pHmeter	2.2	2.5	2.5	2-3	2.5		
Density 15C (Kg/L)	ASTM D4052	1.207	1.2	1.15-1.2	1.192	1.2		1.19
HHV (MJ/Kg)	DIN51900	17.57						
LHV (MJ/Kg)	DIN51900	15.83						
Solids Content (%wt)	Insolubles in Ethanol	0.06						
Ash content (%wt)	ASTM D482	0.0034	<0.1		0.1	0.15	0-0.2	
Pour point	ASTM D97	-30		-30	-30			
Flash point	ASTM D93	48		40-65	40-65	51		
Viscosity (cP) @ 40C	ASTM D445	47.18	40	40-100	40-100		43-1510	
Viscosity 20°C 9mm²/s)	ASTM D445							
Viscosity 50°C (mm²/s)	ASTM D445	9.726						
Carbon (%wt)	ASTM D5291	42.64	40.1	51.1	~52	39.17	54-58	39.4-46.7
Hydrogen (wt%)	ASTM D5291	5.83	7.6	7.3	~6.4	8.04	5.5-7	7.2-7.9
Nitrogen (wt%)	ASTM D5291	0.1	0.1		~0.2	0.05	0-0.2	
Sulphur (%wt)	ASTM	0.01						0.032
Clorine (%wt)	ASTM	0.012						0.2
AlkaliMetals (%wt)	ICP	<0.003						
Oxygen (wt%)		52.1	41.6		~40	52.74	35-40	45.7-52.7

2.3.3 PYROLYSIS OIL UPGRADING

Pyrolysis oil is the product of fast pyrolysis of biomass, a process that allows the decomposition of large organic compounds of biomass such as lignin at medium temperatures in the presence of oxygen. Pyrolysis, that is in essence thermal cracking of biomass, is a well established process for producing bio-oil, the quality of which however is far too poor for direct use as transportation fuel. The product yields and chemical composition of pyrolysis oils depend on the biomass type and size as well as on the operating parameters of the fast pyrolysis. However, a major distinction between pyrolysis oils is based on whether catalyst is employed for the fast pyrolysis reactions or not. Non-catalytic pyrolysis oils have a higher water content than catalytic pyrolysis oils, rendering the downstream upgrading process a more challenging one for the case of the non-catalytic pyrolysis oils.

Untreated pyrolysis oil is a dark brown, free-flowing liquid with about 20-30% water that cannot be easily separated. It is a complex mixture of oxygenated compounds including water solubles (acids, alcohols, ethers) and water insolubles (n-hexane, di-chloor-methane), which is unstable in long-term storage and is not miscible with conventional hydrocarbon-based fuels. It should be noted that due to its nature pyrolysis oil can be employed for the production of a wide range of chemicals and solvents. However, if pyrolysis oil is to be used as a fuel for heating or transportation, it requires upgrading leading to its stabilization and conversion to a conventional hydrocarbon fuel by removing the oxygen through catalytic hydrotreating. For this reason, a lot of research effort is focused on catalytic hydrotreating of pyrolysis oil, as it is a process enabling oxygen removal and conversion of the highly corrosive oxygen compounds into aromatic and paraffinic hydrocarbons.

For non-catalytic pyrolysis oils, the catalytic hydrotreating upgrading process involves contact of pyrolysis oil molecules with hydrogen under pressure and at moderate temperatures (<400°C) over fixed bed catalytic reactors. Single-stage hydrotreating has proved to be difficult, producing a heavy, tar-like product. Dual-stage processing, where mild hydrotreating is followed by more severe hydrotreating has been found to overcome the reactivity of the bio-oil. Overall, the pyrolysis oil is almost completely deoxygenated by a combination of hydro deoxygenation and decarbox-ylation. In fact less than 2% oxygen remains in the treated, stable oil, while water and off-gas are also produced as byproducts. The water phase contains some dissolved organics, while the off-gas contains light hydro-carbons, excess hydrogen, and carbon dioxide. Once the stabilized oil is produced it can be further processed into conventional fuels or sent to a refinery. Table 1 shows the properties of some common catalytic pyrolysis oils according to literature.

Catalytic pyrolysis oils have been reported to getting upgraded via single step hydroprocessing, most of the times utilizing conventional CoMo and NiMo catalysts. During the single step hydroprocessing, the catalytic pyrol-ysis oil feedstock is pumped to high pressure, then mixed with compressed hydrogen and enters the hydroprocessing reactor. In Table 5 the typical oper-ating parameters for single stage hydroprocessing and associated deoxygen-ation achievements are given according to literature [29;33-38].

TABLE 5: Single-stage pyrolysis oil hydroprocessing operating parameters

Catalyst	CoMo [29][33][34][35][36], NiMo [34][35][36], others [37][38]
Temperature (°C)	350-420
Pressure (psig)	1450-2900
LHSV (Hr⁻¹)	0.1-1.2
Deoxygenation (wt%)	78-99.9
Density (kg/l)	0.9-1.03

However, in the case of non-catalytic pyrolysis oils or for achieving bet-ter quality products, multiple-stage hydroprocessing can be employed for upgrading pyrolysis oils. Multiplestage hydroprocessing utilizes at least two different stages of hydroprocessing, which may include hydrotreating or hydrotreating and hydrocracking reactions. In the first stage the cata-lytic hydrotreatment reactor stabilizes the pyrolysis oil by mild hydrotreat-ment over CoMo or NiMo hydrotreating catalyst [32;40-42]. The first stage product is then further processed in the second-stage hydrotreater, which operates at higher temperatures and lower space velocities than the first stage hydrotreater, employing also CoMo or NiMo catalysts within the reactor. The 2nd stage product is separated into an organic-phase product, wastewater, and off-gas streams. In the literature [41], even a 3rd stage hydroprocessing has been used for the heavy fraction (which boils above 350°C) of the 2nd stage product, where hydrocracking reactions take place for converting the heavy product molecules into gasoline and diesel blend components.

TABLE 6: Multiple-step pyrolysis oil hydroprocessing operating parameters

Feed	1st stage	2nd stage	3rd stage
Catalyst	CoMo[32][40],NiMo[32][42], others[39]	CoMo[32][40]NiMo[32][42], others [39]	CoMo[4141]
Temperature (C°)	150-240	225-370	350-427
Pressure (psig)	1000-2000	2015	1280
LHSV (hr^{-1})	0.28-1	0.05-0.14	
Deoxygenation (wt%)		60-98.6	

2.3.4 FISCHER-TROPSCH WAX UPGRADING

Biofuels production via the Fischer-Tropsch technology is a conversion process of solid biomass into liquid fuels (Biomass-To-Liquid or BTL) as it is depicted in Figure 2. More specifically the solid biomass is gasified in the presence of air and the produced biogas rich in CO and H_2 (synthesis gas), after being pretreated to remove coke residues and sulfur compounds, enters the Fischer-Tropsch reactor. The Fischer-Tropsch reactions allow the catalytic conversion of the synthesis gas into a mixture of paraffinic hydrocarbons consisting of light (C_1-C_4), naphtha (C_5-C_{11}), diesel (C_{12}-C_{20}) and heavier hydrocarbons (>C_{20}). Even though the Fischer-Tropsch reactions yields depend on the catalyst and operating parameters employed [43-45], the liquid product (naphtha, diesel and heavier hydrocarbons) yield is high (~95%). The produced synthetic naphtha and diesel fuels can be used similarly to their fossil counterparts. The heavier product however, which is called as Fischer-Tropsch wax, due to its waxy/

paraffinic nature should get upgraded via catalytic hydrocracking to get converted to mid-distillate fuels (naphtha and diesel).

The conversion of Fischer-Tropsch wax into mainly diesel was studied in virtue of the European Project RENEW [46]. During this project Fischer-Tropsch wax with high paraffinic content of C_{20}-C_{45} was converted into a total liquid product consisting of naphtha, kerosene and diesel fractions via catalytic hydrocracking. However the total liquid product content of diesel molecules was the highest and the diesel fraction was further separated and characterized having density of 0.78gr/ml and cetane index of 76 [47]. The schematic of the BTL process with actual images of the feedstock, Fischer-Tropsch wax and synthetic diesel are given in Figure 8.

2.3.5 MICRO-ALGAL OIL CONVERSION TO BIOFUELS

The rapid development of the biofuels production technologies from different biomass types has given rise to the biomass and food markets as it was aforementioned. Besides the use of residual biomass, research and in particular biotechnology has moved forward into seeking alternative biomass production technologies that will offer higher yields per hectare as well as lipids and carbohydrates, which are not part of the human and animal foodchain, avoiding competition between food/feed and energy crops. Targeted research efforts have offered a promising solution by the selection of unicellular microorganisms for the production of biofuels [48][49]. Micro-algae are photosynthetic microorganisms that can produce lipids, proteins and carbohydrates in large amounts over short periods of time.

Micro-algae are currently considered a prominent source of fatty acids, which offers large yields per hectare with various fatty acid foot-prints from each strain. In fact, there are certain strains that offer fatty acids of increased saturation (small content of unsaturated fatty acids) and of smaller carbon-chain length such as *Dunaliellasalina, Chlorella minutissima, Spirulina maxima, Synechococcus sp.*[50] etc. Another advantage of algal oils is that their fatty acid content can be directed to small carbon-chain molecules either genetically or by manipulating the aquaculture conditions such as light source and intensity [51], nitrogen starvation period [52], nutrients and CO_2 feeding profiles [53].

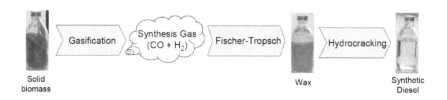

FIGURE 8: Biomass-to-Liquid production of synthetic diesel

Micro-algae and their products formulated the so called 3rd generation biofuels, as they incorporate various characteristics, which render them superior over other biofuels and biomass types. Micro-algae can also be produced in sea water [54] or even waste water, while they are biodegradable and relatively harmless during an eventual spill. Furthermore, their yield per hectare can reach 3785-5678lit, which is 20-700 higher over the conventional energy crops yield (soy, rape and palm). The lipids contained in most micro-algal oils have a similar synthesis with that of soy-bean oil, while they also contain some poly-saturated fatty acids with four double bonds. As a result catalytic hydrotreating of micro-algal oil is the most promising technology for converting it into biofuels.

2.3.6 CO-HYDROPROCESSING

The effectiveness of catalytic hydroprocessing was also explored for co-processing of lipid feedstocks with petroleum fractions as catalytic hydro-processing units are available in almost all refineries. The first co-process-ing study involved experiments of catalytic hydrotreating of sunflower oil mixtures with heavy petroleum fractions aiming to produce high quality diesel [55]. The experiments were conducted in a continuous fixed-bed reactor over a wide range of temperatures 300-450°C employing a typi-cal $NiMo/Al_2O_3$ hydrotreating catalyst. The study was focused on the hy-drogenation of double C-C bonds and the subsequent paraffin formation via the three different reactions routes: decarbonylation, decarboxylation and deoxygenation. Furthermore the large carbon-chain paraffins can also undergo isomerization and cracking leading to the formation of smaller paraffins. This study concluded that the selectivity of products on decarboxylation and decarbonylation is increasing as the temperature and vegetable oil content in the feedstock increase [55].

In a similar study catalytic hydrocracking over sunflower oil and heavy vacuum gas oil mixtures was investigated [56]. The experiments were conducted in a continuous-flow hydroprocessing pilot-plant over a range of temperatures (350-390°) and pressures (70-140bar). Three different hydrocracking catalysts were compared under the same conditions and four different feedstocks were employed, incorporating for 10% and 30%v/v of lipid bio-based feedstock and considering non-pretreated and pretreated sunflower oil as a bio-based feedstock. The results indicated that a prior mild hydrogenation step of sunflower oil is necessary before hydrocracking. Furthermore, conversion was increased with increasing sunflower oil ratio in the feedstock and increasing temperature, while the later decreased diesel selectivity.

The effect of the process parameters and the vegetable oil content of the feedstocks on the yield, physical properties, chemical properties and application properties during co-hydrotreating of sunflower oil and gas-oil mixtures utilizing a typical $NiMo/Al_2O_3$ hydrotreating catalyst was also studied [57]. The experimental results of this study indicated that catalytic co-hydrogenation of gas oil containing sunflower oil in different percentages allowed both vegetable oil conversion reactions (saturation, deoxygenation) and the gas oil quality improvement reactions (hetero atom removal, aromatic reduction). The optimal operating conditions (360-380°C, P=80 bar, LHSV=1.0h-1, H_2/oil=600 Nm^3/m^3 and 15% sunflower oil content of feed) resulted in a final diesel product with favorable properties (e.g. less than 10 wppm sulfur, ~20% aromatics) but poor cold flow properties (CFPP=3°C). The study also indicated that for sunflower content in the feedstock higher than 15% reduced the desulfurization efficiency. Furthermore, the authors also concluded that the presence of sunflower oil in the feedstock has augmented the normal and iso-paraffins content of the final product and as a result has increased the cetane number but degraded the cold flow properties, indicating that an isomerization step is required as an additional step.

The issue of catalyst development suitable for co-hydrotreating and co-hydrocracking of gas-oil and vegetable oil mixtures was recently addressed [10], as there are no commercial hydroprocessing catalysts available for lipid feedstocks. New sulfided $Ni-W/SiO_2-Al_2O_3$ and sulfided $Ni-Mo/Al_2O_3$ catalysts were tested for hydrocracking and hydrotreating of gas- oil and vegetable oil mixtures respectively. The results indicated that the hydrocracking catalyst was more selective for the kerosene hydrocarbons (140–250°C), while the less acidic hydrotreating catalyst was more selective for the diesel hydrocarbons (250–380°C). The study additionally showed that the deoxygenation reactions are more favored over the hydrotreating catalyst, while the decarboxylation and decarbonylation reactions are favored over the hydrocracking catalyst.

2.4 DEMO AND INDUSTRIAL APPLICATIONS

As catalytic hydrotreating of liquid biomass has given promising results, the industrial world has given enough confidence to apply it in pilot and industrial scale. The NesteOil Corporation has developed the NExBTL technology for converting vegetable oil (primarily palm oil) into a renewable diesel also known as "green" diesel (Figure 9). Based on this technology the first catalytic hydrotreatment of vegetable oils unit was constructed in Finland in 2007, within the existing Poorvo refinery of NesteOil, with a capacity of 170 kton/hr. The primary feedstock is palm oil, while it can also process rapeseed oil and even waste cooking oil. The same company has constructed a second unit within the same refinery while it has also planned to construct two new units, one in Singapore and one in Rotterdam, with the capacity of 800 kton/yr each.

The catalytic hydrotreatment technology of 100% waste cooking oil for biodiesel production was developed in the Centre for Research and Technology Hellas (CERTH) in Thessaloniki, Greece [21-24] and later demonstrated via the BIOFUELS-2G project [59], which was co-funded by the European Program LIFE". In this project WCO was collected from associated restaurants and the produced 2nd generation bio-diesel, to be called "white diesel" was employed. For the demonstration of the new technology, 2 tons of "white diesel' were produced via catalytic hydrotreatment of WCO based on the large-scale pilot units available in CERTH. The production process simplified diagram is given in Figure 10. The new fuel will be applied to a garbage truck in a 50-50 mixture with conventional diesel in August 2012, aiming to promote the new technology as it exhibits overall yields exceeding 92% v/v.

In the USA the Dynamic Fuels company [60] has constructed in Baton Rouge a catalytic hydrotreating unit dedicated to oils and animal fats with 285 Mlit capacity. The unit employs the Syntroleum technology based on Fischer-Tropsch for the production of synthetic 2nd generation Biodiesel while it also produces bio-naphtha and bio-LPG. The Bio-Synfining technology of Syntroleum converts the triglycerides of fats and oils into n- and iso-paraffins via catalytic hydrogenation, thermal cracking and isomerization as it is applied in the Fischer- Tropsch wax upgrading to renewable diesel (R-2) and renewable jet (R-8) fuel.

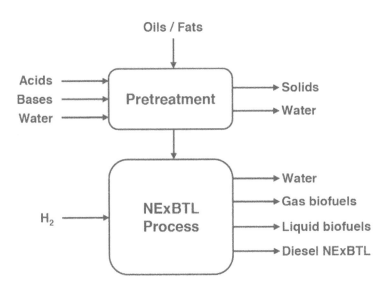

FIGURE 9: NExBTL catalytic hydrotreating of oils/fats technology for biodiesel production [58]

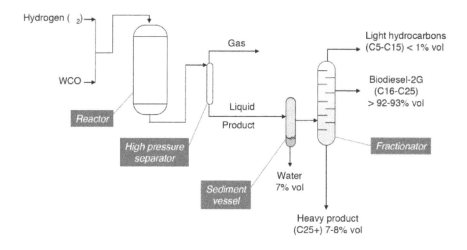

FIGURE 10: Pilot-scale demo application of WCO catalytic hydrotreatment during the BIOFUELS-2G project [59]

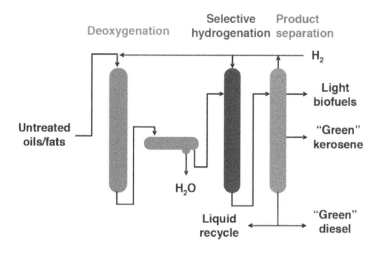

FIGURE 11: Vegetable oil and animal fats conversion technology to renewable fuels of UOP [61]

TABLE 7: Pilot flights with biofuels [62]

Airline	Aircraft	Partners	Biofuel (lipid sources)	Blend*
Virgin Atlantic	B747-400	Boeing, GE Aviation	FAME (coconut & palm)	20%
Air New Zealand	B747-400	Boeing, Rolls-Royce	HRJ (Jatropha)	50%
Contintental Airlines	B737-800	Boeing, GE Aviation, CFM, Honeywell UOP	HRJ (Jatropha&algea)	50%
JAL	B747-300	Boeing, Pratt&Whitney, Honeywell UOP	HRJ (Camelina, Jatropha& algae)	50%
KLM	B747-400	GE, Honeywell UOP	HRJ (Camelina)	50%
TAM	A320	Airbus, CFM	HRJ (Jatropha)	50%

2.5 FUTURE PERSPECTIVES

Catalytic hydrotreating of liquid biomass is continuously gaining ground as the most effective technology for liquid biomass conversion to both ground- and air-transportation fuels. The UOP company of Honeywell, via the technology it has developed for catalytic hydrotreating of liquid biomass (Figure 11), has announced imminent collaboration with oil and airline companies such as Petrochina, Air China and Boeing for the demonstration of the sustainable air-transport in China. This initiative will lead a strategic collaboration between the National Energy Agency of china with the Commerce and Development Agency of USA leading to the development of the new biofuels market in China.

In the EU airline companies collaborate with universities, research centers and biofuels companies in order to confront their extensive contribution to CO_2 emissions. Since 2008 most airline companies promote the use of biofuels in selected flights as shown in Table 7 [62]. As it is obvious most pilot flights have taken place with Hydrotreated Renewable Jet (HRJ), which is kerosene/jet produced via catalytic hydrotreatment of liquid biomass. Moreover, Lufthansa has also completed a 6-month exploration program of employing HRJ in a 50/50 mixture with fossil kerosene in one of the 4 cylinders of a plane employed for the flight between Hamburg-Frankfurt-Hamburg with excellent results [63].

Besides the future applications for air-transportation, the automotive industry is also exhibiting increased interest for the broad use of biofuels resulting from catalytic hydrotreatment of liquid biomass. In fact these paraffinic biofuels can be employed in higher than 7%v/v blending ratio (which is the maximum limit for FAME) as they exhibit high cetane number and have significant oxidation stability [64]

The highest interest is exhibited by oil companies around the catalytic hydrotreatment of liquid biomass technology for the production of biofuels and particularly to its application to oil from micro-algae. ExxonMobil has invested 600M$ in the Synthetic Genomics company of the pioneer scientist Craig Ventner aiming to research of converting micro-algae to biofuels with minimal cost. BP has also invested 10M$ for collaboration with Martek for the production of biofuels from micro-algae for air-, train-, ground- and marine transportation applications.

2.6 CONCLUSION

Catalytic hydrotreatment of liquid biomass is the only proven technology that can overcome its limitations as a feedstock for fuel production (low H/C ratio, high oxygen and water content). Even though it has recently started to be investigated as an alternative technology for biofuels production, it fastly gains ground due to the encouraging experimental results and successful pilot/demo and industrial applications. Catalytic hydrotreatment of liquid biomass leads to a wide range of new alterative fuels including bio-naphtha, bio-jet and biodiesel, are paraffinic in nature and as a result exhibiting high heating values, increased oxidation stability and negligible acidity and corrosivity. As a result it is not over-optimistic to claim that this technology will broaden the biofuels market into scales capable to actually mitigate the climate change problems.

REFERENCES

1.   Meher LC, Vidya SD, Naik SN. Technical aspects of biodiesel production by trans-
     esterification—a review. Renewable and Sustainable Energy Reviews 2006;10
     248–268.

2. Naik SN., Goud VV, Rout OJ, Dalai AK. Production of first and second generation biofuels: A comprehensive review. Renewable and sustainable Energy Reviews 2010;14 578-597.
3. Gary JH, Handwerk GE. Petroleum Refining, Technology and Economics, 4th ed. New York: Marcel Dekker; 2001.
4. Furimsky E. Selection of catalysts and reactors for hydroprocessing. Applied Catalysis A: General 1998;171(2) 177-206.
5. Birchem T. Latest Improvements in ACETM Catalysts Technology for ULSD Production & Deep Cetane Increase. 5th ERTC Annual Meeting: conference proceedings, November 2010, Istanbul, Turkey.
6. Donnis B, Egeberg RG, Blom P, Knudsen KG. Hydroprocessing of bio-oils and oxygenates to hydrocarbons. Understanding the reaction routes. Top. Catal. 2009; 52 229–240.
7. Boda L, Onyestyák G, Solt H, Lónyi F, Valyon J, Thernesz A. Catalytic hydroconversion of tricaprylin and caprylic acid as model reactionfor biofuel production from triglycerides. Applied Catalysis A: General 2010;374 158–169.
8. Bezergianni S, Kalogianni A, Dimitriadis A, Catalyst Evaluation for Waste Cooking Oil Hydroprocessing, Fuel 2012;93 638-647.
9. Funimsky E. Catalytic Hydrodeoxygenation. Applied Catalysis A: General 2000;199147-190.
10. Tiwari R, Rana BS, Kumar R, Verma D, Kumar R, Joshi RK, Garg MO, Sinha AK. Hydrotreating and hydrocracking catalysts for processing of waste soya-oil and refinery-oil mixtures. Catalysis Communications 2011;12 559-562.
11. Lavrenov AV, Bogdanets EN, Chumachenko YA, Likholobov VA. Catalytic processes for the production of hydrocarbon biofuels from oil and fatty raw materials: Contemporary approaches. Catalysis in Industry 2011;3(3) 250-259.
12. Bulushev DA, Ross JRH. Catalysis for conversion of biomass to fuels via pyrolysis and gasification: A review. Catalysis Today 2011; 171(1) 1-13.
13. Bezergianni S, Dimitriadis A, Kalogianni A., Pilavachi PA. Hydrotreating of waste cooking oil for biodiesel production. Part I: Effect of temperature on product yields and heteroatom removal. Bioresource Technology 2010;101 6651–6656.
14. Tyson KS. Biodiesel Handling and Use Guidelines. National Renewable Energy Laboratory 2001, NREL/TP-580-30004
15. Winayanuwattikun P, Kaewpiboon C, Piriyakananon K, Tantong S, Thakernkarnkit W, Chulalaksananukul W, Yongvanich T. Potential plant oil feedstock for lipase-catalyzed biodiesel production in Thailand. Biomass & Bioenergy 2008;321279–1286.
16. Da Rocha Filho GN, Brodzki D, Djega-Mariadassou G. Formation of alkanes alkylcykloalkanes and alkylbenzenes during the catalytic hydrocracking of vegetable oils. Fuel 1993;72 543–549.
17. Gusmao J, Brodzki D, Djega-Mariadassou G, Frety R. Utilization of vegetable oils as an alternative source for diesel-type fuel: Hydrocracking on reduced Ni/ SiO2 and sulphided Ni–Mo/c-Al2O3. Cat Today 1989;5 533–544.
18. Simacek P, Kubicka D, Sebor G, Pospisil M. Fuel properties of hydroprocessed rapeseed oil. Fuel 2010;89611–615.
19. Simacek P, Kubicka D, Sebor G, Pospisil M. Hydroprocessed rapeseed oil as a source of hydrocarbon-based Biodiesel. Fuel 2009;88456–60.

20. Bezergianni S, Kalogianni A. Hydrocracking of used cooking oil for biofuels production. Bioresource Technology 2009;100(17) 3927-3932.
21. Bezergianni S, Dimitriadis A, Kalogianni A, Pilavachi PA. Hydrotreating of waste cooking oil for biodiesel production. Part I: Effect of temperature on product yields and heteroatom removal. Bioresource Technology 2010;101 6651-6656.
22. Bezergianni S, Dimitriadis A, Kalogianni A, Knudsen KG. Toward Hydrotreating of Waste Cooking Oil for Biodiesel Production. Effect of Pressure, H2/Oil Ratio, and Liquid Hourly Space Velocity. Industrial Engineering Chemistry Research 2011;50(7) 3874-3879.
23. Bezergianni S, Dimitriadis A, Sfetsas T, Kalogianni A. Hydrotreating of waste cooking oil for biodiesel production. Part II: Effect of temperature on hydrocarbon composition. Bioresource Technology 2010;101 7658-7660.
24. Bezergianni S, Dimitriadis A, Voutetakis S. Catalytic Hydrotreating of Waste Cooking Oil for White Diesel Production. Proceedings of First International Congress on Catalysis for Biorefineries (CatBior), October 2-5,Torremolinos-Málaga, Spain; 2011
25. Karonis D, Bezergianni S, Chilari D, Kelesidis E. Maximizing the yield of biodiesel from waste cooking oil hydroprocessing by cut-point optimization. Proceedings Am. Chem. Soc., Div. Fuel Chem. 2010; 55(1)
26. DynamotiveBioOil Information Booklet 2009, http://ingenieria.read.net/blog/PIB-BioOil.pdf (last visited August 6, 2012)
27. Venderbosch RH, Ardiyanti AR, Wildschut J, Oasmaa A, Heeres HJ. Stabilization of biomass-derived pyrolysis oils. J Chem.Technol.Bioetchnol. 2010; 85 674-686.
28. Wildschut J, Mahfud FH, Venderbosch RH, Heeres HJ. Hydrotreatment of fast pyrolysis Oil Using Heterogeneous Noble-Metal Catalysts. Ind. Eng. Chem. Res. 2009;48 10324-10334.
29. Wildschut J. Pyrolysis Oil Upgrading to Transportation Fuels by Catalytic Hydrotreatment. PhD Thesis. Rijksuniversiteit Groningen; 2009.
30. Baldauf W, Balfanz U, Rupp M. Upgrading of Flash Pyrolysis Oil and Utilization in Refineries. Biomass and Bioenergy 1994;7(1-6) 237-244.
31. Czernik S, Bridgwater AV. Overview of Applications of Biomass Fast Pyrolysis Oil. Energy& Fuels 2004;18 590-598.
32. Elliott DC, Neuenschwander GG. Liquid fuels by low severity hydrotreatment of biocrude. Developments in Thermochemical Biomass Conversion, Vol. 1, pp. 611-621, A. V. Bridgwater and D. G. B. Boocock, eds., Blackie Academic & Professional, London: 1996
33. Baker EG, Elliott DC. Catalytic Hydrotreating of Biomass-Derived Oils. Pacific Northwest Laboratory 1988.
34. BorjeGevert S, Anderson PV, Sandqvist SP. Hydroprocessing of directly liquefied biomass with large-pore catalysts. Energy & Fuels 1990;4 78-81.
35. Elliott DC, Baker EG. Upgrading Biomass Liquefaction Products through hydrodeoxygenation. Biotechnol.Bioeng.Symp. 1984, suppl. 14, 159.
36. Samolada MC, Baldauf W, Vasalos IA. Production of a Bio-Gasoline by Ugrading Biomass Flash Pyrolysis Liquids via Hydrogen Processing and Catalytic Cracking. Fuel 1998;77 1667.

37. de Miguel Mercader F, Groeneveld MJ, Kersten SRA, Way NWY, Schaverien CJ, Hogendoorn JA. Production of advanced biofuels: Co-processing of upgraded pyrolysis oil instandard refinery units. Applied Catalysis B: Environmental 2010;96 57–66.

38. Oasmaa A, Kuoppala E, Ardiyanti A, Venderbosch RH, Heeres HJ. Characterization of Hydrotreated Fast Pyrolysis Liquids. Energy Fuels 2010;24 5264–5272.

39. Venderbosch RH, Heeres HJ. Pyrolysis Oil Stabilisation by Catalytic Hydrotreatment. Biofuel's Engineering Process Technology.

40. Jones SB, Holladay JE, Valkenburg C, Stevens DJ, Walton C, Kinchin C, Elliott DC, Czernik S. Production of Gasoline and Diesel from Biomass via Fast Pyrolysis, Hydrotreating and Hydrocracking: A Design Case. Prepared for the U.S. Department of Energyunder Contract DE-AC05-76RL01830. February 2009.

41. Pindoria RV, Megaritis A, Herod AA, Kandiyoti R. A two-stage fixed-bed reactor for direct hydrotreatment of volatiles from the hydropyrolysis of biomass: effect of catalyst temperature, pressure and catalyst ageing timeon product characteristics. Fuel 1998;77(15) 1715–1726.

42. Conti L, Scano G, Boufala J, Mascia S. Experiments of Bio-oil Hydrotreating in a Continuous Bench-Scale Plant. In: Bridgwater, A.V.; Hogan, E.N., editors. Bio-Oil Production and Utilization. CPL Press, Newbury Berks, 1996, 198

43. Sharma RK, Bakhshi NN. Catalytic Upgrading of Pyrolysis Oil. Energy & Fuels 1993;7 306-314.

44. Subiranas M, Schaub A. Combining Fischer-Tropsch (FT) and hydrocarbon reactions under FT reaction conditions - Catalyst and reactor studies with Co or Fe and Pt/ ZSM-5, International Journal of Chemical Reactor Engineering 2007; 5(A78).

45. Rosyadi E, Priyanto U, Suprapto, Roesyadi A, Nurunnabi M, Hanaoka T, Miyazawa T, Sakanishi K, Biofuel production by hydrocracking of biomass FT wax over NiMo/ Al2O3-SiO2 catalyst, Journal of the Japan Institute of Energy 2011;90(12) 1171-1176.

46. www.renew-fuel.com (accessed 8 August 2012)

47. Lappas AA, Voutetakis SS, Drakaki N, Papapetrou M., Vasalos IA. Production of Transportation Biofuels through Mild-Hydrocracking and catalytic cracking of waxes produced from Biomass to Liquids (BTL) Process. Proceedings of the 14thEuropean Biomass Conference, Paris France; 2005

48. Posten C, Schaub G. Microalgae and terrestrial biomass as source for fuels—a process view. J.Biotechnol. 2009;142 64–69.

49. Schenk PM, Thomas-Hall SR, Stephens E, Marx UC, Mussgnug JH, Posten C, Kruse O, Hankamer B. Second Generation Biofuels: High-Efficiency Microalgaefor Biodiesel Production. Bioenerg Res 2008;120-43

50. Patil V, Källqvist T, Olsen E, Vogt G, Gislerød HR. Fatty acid composition of 12 microalgae for possible usein aquaculture feed. Aquaculture International 2007;151-9. http://dx.doi.org/10.5772/52649325

51. Kachroo D, Jolly SMS, Ramamurthy V. Modulation of unsaturated fatty acids content in algae Spirulinaplatensis and Chlorella minutissima in response to herbicide SAN9785. Electronic Journal of Biotechnology 2006;9(4)386-390.

52. Chen M, Tang H, Ma H, Holland TC, Ng KYS, Salley SO. Effect of nutrients on growth and lipid accumulation in the green algae Dunaliellatertiolecta, Bioresource Technology 2011;1021649-1655.

53. Yoo C, Jun SY, Lee JY, Ahn CY, Oh HM. Selection of Microalgae for Lipid Production under High Levels Carbon Dioxide. Bioresource Technology 2010;101 S71-S74.

54. Popovich CA, Damiani C, Constenla D, Martínez AM, Freije H, GiovanardiM, Pancaldi S, Leonardi PI. Neochlorisoleoabundans grown in enriched natural seawater for Biodiesel feedstock: Evaluation of its growth and biochemical composition. Bioresource Technology 2012;114 287–293.

55. Huber GW, O'Connor P, Corma A. Processing biomass in conventional oil refineries: production of high quality diesel by hydrotreating vegetable oils in heavy vacuum oil mixtures. Applied Catalysis A: General 2007;329120–129.

56. Bezergianni S, Kalogianni A, Vasalos IA. Hydrocracking of vacuum gas oil-vegetable oil mixtures for biofuels production. Bioresource Technology 2009;1003036–3042.

57. Tóth C, Baladincz P, Kovács S, Hancsók J. Producing Diesel Fuel by Co-Hydrogenation of Vegetable Oil with Gas Oil. Chemical Engineering Transactions 2010;21 1219-1224.

58. Neste Oil Corporation.NexBTL diesel. http://www.nesteoil.com/default.asp?path=1,41,11991,12243,12335 (accessed 15 July 2012).

59. BIOFUELS-2G project. http://www.biofuels2g.gr(accessed 20 July 2012).

60. Dynamic Fuels. http://www.dynamicfuelsllc.com/ (accessed 18 July 2012).

61. Honneywell's UOP. http://www.uop.com (accessed 5 July 2012).

62. Steele P. Powering the Future of Flight. In Proceedings of the World Biofuels Markets 2011, March 22-24, Rotterdam, The Netherlands.

63. Buse J. Commercialization of Aviation Biofuels. In Proceedings of the World Biofuels Markets 2011, March 22-24, Rotterdam, The Netherlands.

64. Lahaussois D, Desaeger M. 10% Renewable Energy in transport by 2020: a car manufacturer's perspective. In Proceedings of the World Biofuels Markets 2011, March 22-24, Rotterdam, The Netherlands.

CHAPTER 3

Analytical Approaches in the Catalytic Transformation of Biomass: What Needs to be Analyzed and Why?

DMITRY MURZIN AND BJARNE HOLMBOM

3.1 INTRODUCTION

Today, the use of biomass is considered a promising way to diminish negative environmental impact. Moreover, in some future scenarios, renewable raw materials are thought to be able to replace finite mineral-oil-based raw materials before 2050 [1]. This means that new synthetic routes, which should desirably adhere to the principles of green chemistry [2], need to be developed for the production of chemicals.

Lignocellulosic biomass, as a renewable source of energy and chemicals, has attracted a lot of attention recently [3-10]. Wood biomass consists of cellulose (40–50%), lignin (3–10%), hemicelluloses (15–30%) and a variety of extractives (1–10%). Cellulose is a linear polymer of D-glucopyranose and can contain up to 10,000 units ($C_6H_{10}O_5$), connected by glycosidic ether bonds, while the molecular mass for hemicelluloses is

Murzin D and Holmbom B (2013). "Analytical Approaches in the Catalytic Transformation of Biomass: What Needs to be Analyzed and Why?" in Catalysis for the Conversion of Biomass and Its Derivatives, *Behrens M and Datye AK (Eds.), ISBN: 9783844242829, Reprinted with permisison from the authors.*

lower. Hemicelluloses have a more heterogeneous structure than cellulose, consisting mainly of five-carbon (xylose, arabinose) and six-carbon sugars (galactose, glucose and mannose). Contrary to cellulose lignin is a co-niferyl alcohol polymer with coumaryl, coniferyl and sinapyl alcohols as monomers, which are heavily cross-linked, leading to complex structures of large lignin molecules [11].

Chemical treatment of lignocellulosic biomass in general, and wood in particular, can have several targets. One of the options is delignifi-cation of the biomass leading to cellulose and some residual hemicel-luloses, which are further applied in the production of paper or board, or derivatives of cellulose. Thermal (or catalytic) treatment of biomass, e.g., thermal or catalytic pyrolysis, is a route to bio-based synthesis gas and biofuels [12]. Depolymerization results in the formation of low-molecular-mass components (sugars, phenols, furfural, various aromatic and aliphatic hydrocarbons, etc.), e.g., unique building blocks for further chemical synthesis.

Wood biomass contains many valuable raw materials for producing fine and specialty chemicals (Figure 1). These raw materials are carbohy-drates, fatty acids, terpenoids and polyphenols, such as stilbenes, lignans, flavonoids and tannins. Some of them can be exuded directly from living trees, while others are extracted and purified via chemical methods.

In this context, applications of catalytic reagents, which are superior to stoichiometric reagents producing stoichiometric amounts of wastes, are worth mentioning. Well-known benefits in using heterogeneous catalysts are associated with easy catalyst separation, regeneration and reuse, as well as relatively low prices compared to homogeneous catalysts. The re-search regarding catalytic transformations of different wood-derived com-pounds is currently very active [13].

Because of the complexity associated with the processing of biomass per se or the transformation of biomass-derived chemicals, in-depth chemical analysis of all components and their reactions is difficult to perform. Therefore, most analytical methods will be a result of a com-promise between information depth and available resources. It is also obvious that in industrial processes only a limited number of rather fast analytical methods could be utilized since a large number of samples should be processed.

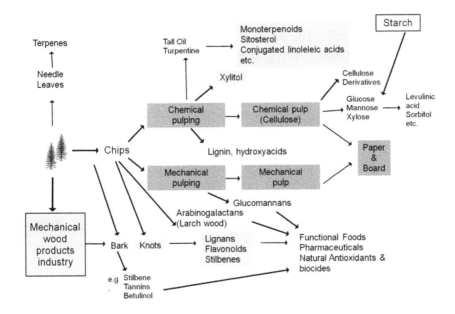

FIGURE 1: Chemical by-products from the forest industry

To have in-depth and molecular-level understanding of the chemical reactions occurring during the transformation of biomass not only advanced analytical methods are required, but additionally, a broad spectrum of these methods needs to be applied. Let us consider, for example, the catalytic conversion of cellulose [14-17] in the presence of hydrogen leading to sugar alcohols. During such a depolymerization reaction not only the concentration of carbohydrates and other products in the liquid phase should be measured, but also the crystallinity of cellulose, its morphology, molecular mass distribution and presence of sugar oligomers. The analysis is even more complicated if in this reaction wood is used directly instead of cellulose.

Analytical techniques have made a tremendous progress in recent years giving a possibility to utilize a wide range of modern instrumental methods, including advanced chromatography, microscopy and spectroscopy. It is apparently clear that all the methods currently available cannot be treated in this review, thus a rational selection of them was done by the authors based on their experience, with an understanding that it might not cover all the analytical methods presently utilized in catalytic transformations of biomass-derived chemicals, but focuses mostly on chromatography.

3.2 ANALYTICAL OBJECTIVES

Any planning of analytical procedures should be based on the goals and scope of the study. The following critical steps in an analytical process can be listed: problem definition and formulation of analytical objectives; setup of an analytical plan; sampling; sample transport and storage; sample pretreatment; analytical determination; data calculation; evaluation of results to see if the objectives are achieved.

It is apparently clear from this list that the actual analytical determination is just one step among the others and sometimes could not even be the crucial one. Moreover, preparation of the samples, pretreatment and evaluation of data could be more demanding or at least time-consuming. Since in catalytic transformation of lignocellulosic biomass often wood or various streams from pulping are used as raw materials, a special attention should be devoted to sampling. Inappropriate sampling could undermine the value of the whole study, therefore it should be carefully planned. Sampling and sample storage is important since samples may be altered or destroyed due to temperature, light, presence of oxygen, humidity, enzymes or microbes (bacteria, fungi, etc.). For instance, enzymatic and microbiological attack can happen for samples of fresh wood, wet pulp and paper, sludge, process waters and effluents, while polyunsaturated extractives like abietic acid could be subjected to oxidation.

3.3 BASIC ANALYTICAL METHODS

3..3.1 CHROMATOGRAPHY

Chromatographic and spectroscopic methods are widely used today for analytical purposes. Chromatographic techniques are applied not only for off-line analysis, but also for the on-line determination of minute amounts, as well as large-scale preparative separations. In fact, not only monomers, but also polymers and oligomers can be separated by chromatography, although in the former case it is essentially a group separation. There are several forms of chromatography using different mobile and stationary phases, with the two main forms of instrumental chromatography being liquid (LC) and gas chromatography (GC). According to IUPAC definition, chromatography is a physical method of separation in which the components to be separated are distributed between two phases, one of which is stationary, while the other (mobile) moves in a definite direction.

3.3.2 GAS CHROMATOGRAPHY

When relating gas chromatography to catalytic transformations of biomass, it can be stated that GC (Figure 2) provides qualitative and quantitative determination of organic components such as extractives, hemicel-

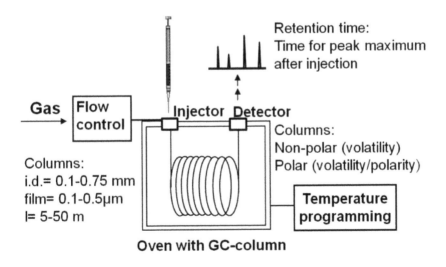

FIGURE 2: The schematic of gas chromatography

lulose building blocks, organic acids, etc. The derivatized and vaporized products are introduced to the column for separation and identified in a detector, whose response is recorded as a chromatogram. Capillary columns made of fused silica with a stationary phase as a thin film of liquid or gum polymer on the inside of the tube are mainly used. The most commonly utilized stationary phases are siloxane polymer gums with different substituents providing different polarity. The polymers are usually cross-linked in the column by photolytic or free-radical reactions, bringing strength to the polymer films. Wall-coated open-tubular columns with a liquid phase coated directly on the inner walls, as well as support-coated open-tubular columns are applied. In the latter case a stationary phase is coated on fine particles deposited on the inner walls. Among non-polar columns, HP-1, DB-1, etc., based on dimethyl, polysiloxane could be mentioned. HP-5 with 95% dimethyl polysiloxane and 5% phenyl groups is slightly more polar. Still more polar columns employ polyoxyethylene or polyester liquid phases.

Capillary columns are available in a wide range of internal diameters, lengths and liquid film thicknesses (Figure 2). Although longer columns provide better separation, they have an increased analysis time which is usually undesired. In addition, longer columns lead to higher pressure and thus to problems with the injection. Columns with thicker films have higher capacity, but usually require higher temperature, while thin-film columns are suited for large molecules with low volatility. In principle, analysis of components with up to 60 carbon atoms is possible.

Different types of injection systems are used in GC. Split mode, where the injected material after evaporation is split between the column and an outlet, affords rapid volatilization and homogeneous mixing with the carrier gas. Most of the sample will pass out through the split vent and only a small proportion will flow into the column. Splitless systems provide a more reliable quantification allowing analysis of even such high-molecular mass compounds as triglycerides and steryl esters. Flame ionization detectors, which are of destructive nature, have high sensitivity to hydrocarbons, but are not able to detect water. On-line coupling of capillary columns with mass spectrometers is routine nowadays and enables convenient structure identification.

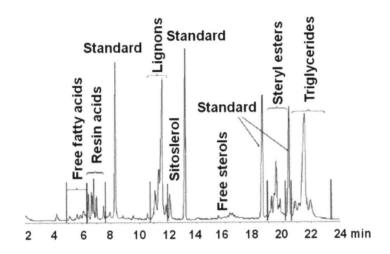

FIGURE 3: Example of a gas chromatogram on a short column with four standards added to a sample (spruce wood extract) [18]

An important but sometimes forgotten issue is the fact that the sensitivity for different compounds is varying for a detector; thus, different peak areas are in proportion to the weight concentration. Knowledge of response factors is therefore necessary and calibration for components especially with various functional groups should be properly done. Commonly, internal standard compounds are applied, e.g., compounds which are not present in the sample itself are purposely added. Chemically they should be similar to the sample compounds with close retention time, however, with no peak overlapping (Figure 3).

In addition to such advantages of GC as accurate quantification based on internal standards, a possibility to be combined with a mass spectrometer and complete automation regarding injection and analytical runs, the very high resolution should also be mentioned. On the other hand only molecules up to about 1000 mass units can be analyzed, as they should be stable at high temperatures. Therefore, sometimes samples should be processed before the analysis. The last point is important for polar compounds, like for example acids, which should be derivatized. GC and GC-MS analysis in the vapour phase require volatile derivatives that do not adsorb onto the column wall. Different derivatizations for different substances are recommended, e.g., silylation or methylation for extractives, methanolysis and silylation for carbohydrates. Silyl derivatives of R-O-Si(CH$_3$)$_3$ type containing a trimethylsilyl group (TMS) are formed by the displacement of the active proton in -OH, -NH and -SH groups. Thus, protic sites are blocked, which decreases dipole-dipole interactions and increases volatility. Common silylation reagents are listed in Figure 4.

Methylation relies on the following reactions: utilization of diazometane (CH$_2$N$_2$): R-COOH + CH$_2$N$_2$ = RCOOMe + N$_2$; acid-catalyzed esterification: R-COOH + R′OH => RCOOR′, as well as on-column esterification using tetra- methyl ammonium salts R-COOH + N+Me$_4$OH- => RCOOMe.

One of the variants of GC is associated with coupling pyrolysis to it (Figure 5). In this arrangement the sample is thermally degraded in an inert atmosphere. The degradation products are introduced to GC or GC-MS for separation and identification allowing qualitative and quantitative determination of semi-volatile and non-volatile components, such as extractives, polymers, paper chemicals, and lignin, etc.

Reagent	Abbreviation	
N, O-Bis-(trimethylsilyl)-acetamide	BSA	CH$_3$—C=NSi(CH$_3$)$_3$ Osi(CH$_3$)$_3$
Hexamethyldisilazane	HMDS	(CH$_3$)$_3$SiNHSi(CH$_3$)$_3$
Trimethylchlorosilane	TMCS	(CH$_3$)$_3$SiCl
Trimethylsilylimidazole	TMSI	(H$_3$C)$_3$Si—N
N,O-Bis-(trimethylsilyl)- trifluoroacetamide	BSTFA	CF$_3$—C=N—Si(CH$_3$)$_3$ Osi(CH$_3$)$_3$

FIGURE 4: Silylation agents

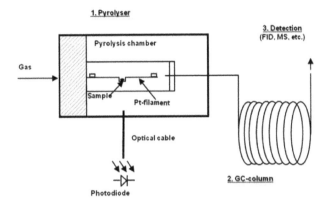

FIGURE 5: Pyrolysis GC

3.3.3 LIQUID CHROMATOGRAPHY

These chromatographic methods use liquids such as water or organic solvents as the mobile phase. Silica or organic polymers as well as anion-exchange resins are used as stationary phase. Separation is performed either at atmospheric pressure or at high pressure generated by pumps. The last version is often called high-performance liquid chromatography (HPLC) with solvent velocity controlled by high-pressure pumps, giving a constant flow rate of the solvents. Solvents are used not only as single solvents but they can also be mixed in programmed proportions. In fact, even gradient elution could be applied with increasing amounts of one solvent added to another, creating a continuous gradient and allowing a sufficiently rapid elution of all components.

The most commonly used columns contain small silica particles (3–10 μm) coated with a nonpolar monomolecular layer.

For lipophilic (low-polar) compounds the mobile phase is an organic solvent, while reversed phase HPLC employs mixtures of water and acetonitrile or water and methanol as eluents and is applied for non-ionized compounds soluble in polar solvents. As examples, such columns (Figure

6) could be mentioned as Agilent Zorbax SB-Aq (4.6×250 mm, 5 µm) allowing the use of highly aqueous mobile phases working in a pH range from 1 to 8 and affording reproducible retention and resolution for polar compounds. Another example is HypercarbTM (4.6×100 mm, 5 µm) with 100% porous graphitic carbon as a stationary phase, which operates in the pH range 0–14 and can resolve highly polar compounds with closely related structures (e.g., geometric isomers, diastereomers, oligosaccharides). CarboPac PA1 (polymer based) column can be used in mono-, oligo- and polysaccharide analysis by high-performance anion-exchange chromatography combined at high pH with pulsed amperometric detection.

UV-Vis (Figure 7) and diode-array detectors enabling recording of UV-Vis spectra, for example every second, are common nowadays. They can be used for the analysis of conjugated and aromatic compounds, such as phenols. Another popular detector is based on refractive index (RI) monitoring and is well suited, for example, for carbohydrates. High-performance anion-exchange chromatography with pulsed amperometric detection is a common technique for analyzing sugars in wood and pulp hydrolysates.

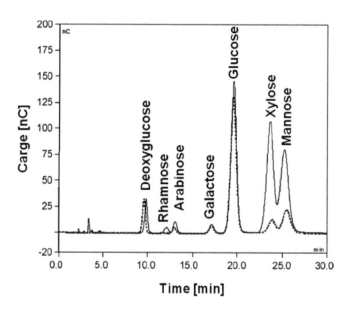

FIGURE 6: Separation of acids and sugars by HPLC using CarboPac PA1 [19]

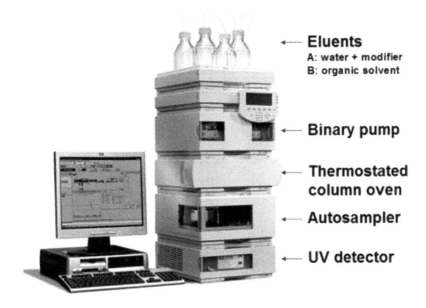

FIGURE 7: A view of LC-UV

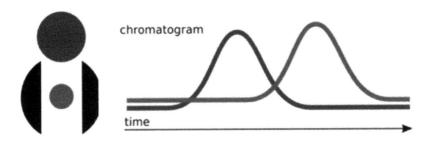

FIGURE 8: Size-exclusion chromatography

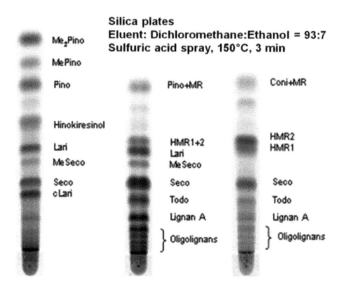

FIGURE 9: TLC of ethanol extracts of knots from: *Araucaria angustifolia* (left), *Abies alba* (center), *Picea abies* trees (right) [21]

Another important form of HPLC is size-exclusion chromatography (Figure 8), which is widely applied for the determination of molecular-mass distributions of dissolved lignin and hemicelluloses, and even for cellulose dissolved in ionic liquids. The same method can be used for the analysis of extractives and their derivatives, for instance dimers and tri-mers of fatty acids [20]. In SEC, solutes in the mobile phase (for example THF) are separated according to their molecular size. Smaller molecules penetrate far into the porous column packing material and thus elute later than larger ones.

The non-destructive character as well as the absence of derivatization could be mentioned among the advantages of LC. This technique can han-dle both small and large amounts and it can be used also for preparative isolation of compounds from mixtures. Contrary to GC there are almost no, or at least much fewer, limitations in terms of the molecular size. In addition, LC can be combined with mass spectrometry, once again with-out derivatization. Thermally unstable and polar compounds can thus be analyzed as such, and the molecular mass in triple quadrupole or ion-trap LC-MS can be up to m/z 3000, while time-of-flight versions allow even up to 16,000.

LC-MS provides better sensitivity and selectivity than GC-MS and is excellent for the quantification of selected substances in complex mixtures. On the other hand, this technique is not very suitable for rapid and reliable identification of unknown compounds mainly because fragmentation is sparse as the conditions of ionization are mild. Furthermore, spectra libraries enabling identification are not available. Other shortcomings of LC-MS are the rather low sensitivity of the detectors for certain compounds. Moreover, it may be difficult to obtain constant pressure, which in turn influences retention; clean, degassed solvents are needed and, finally, it might be challenging to find the optimum solvent mixture.

Nevertheless, there is a large potential in the application of LC-MS toward analysis of oligosaccharides, lignans and oligolignans, flavonoids, stilbenes and tannins, and even fragments of lignin [21].

One form of LC, which is still used in organic synthesis and was popular until the 1960s in the analysis of monosaccharides obtained by hydrolysis of wood, is the so-called planar chromatography or thin-layer chromatography (TLC), where the separation is done in paper sheets or on particle layers deposited on glass, plastic or aluminium plates. Although these times of analysis of carbohydrates are long gone, TLC is an excellent technique for small scale preparative separation of fractions to be further analyzed by GC or LC. During analysis an eluent and the analytes rise in the stationary phase due to capillary forces. The analytes are separated according to their affinity to the stationary phase, which is most commonly silica (Figure 9).

3.3.4 SPECTROMETRIC METHODS

Besides chromatography a wide variety of other techniques are available, such as capillary electrophoresis (CE), Infra Red spectrometry (IR), Nuclear Magnetic Resonance (NMR), Raman, Near Infra Red Spectrometry (NIR) and Ultra Violet-Visual Light Spectrometry (UV-Vis). Electrophoresis is a separation technique based on the differential transportation of charged species in an electric field through a conductive medium. Capillary electrophoresis (CE) was designed to separate species depending on their size to charge ratio in the interior of a small capillary filled with an electrolyte and can be used for analyzing oligosaccharide and monosaccharide reaction products. In the current review we focus mainly on chromatographic methods although the spectrometric methods listed above are certainly of great importance. For instance, UV spectrometry can be used for the determination of lignins in solutions. Colorimetric methods based on selective complexation with special reagents, which can be determined by spectrometric measurements in the UV-Vis range, are applied for the determination of metal ions, hemicelluloses and pectins. IR is a possibility to identify such functional groups as hydroxyls, carbonyls, carboxyls and amines.

For example, the analysis of products in rapeseed oil hydrogenation was conducted by IR [22]. However, IR spectra of large biomolecules are complex, moreover spectra of component mixtures could be difficult to interpret. An advantage of Raman spectroscopy for the transformation of biomass occurring often in water solutions is the easy detection of double and triple carbon-carbon bonds while the adsorption of water is weak. Thus, in contrast to FTIR, wet pulp and wood samples can be analyzed with signals related to extractives, lignin and carbon hydrogen bonds of the polysaccharides, while in FTIR signals of the hydroxyl groups of wood polysaccharides are dominating. NMR is an important method as it provides structural information about complex molecules, therefore it is frequently used for structural analysis of lignins and even hemicelluloses. Crystalline cellulose requires the application of solid-state NMR, as utilized for instance recently in the hydrolytic hydrogenation of cellulose [17].

3.3.5 SELECTION OF ANALYTICAL METHODS

Summarizing shortly the methods described above, it can be stated that the choice of analytical methods in general depends on sample characteristics, matrix complexity, the aim of the analysis, accessible equipment and the amount of resources available.

For instance in order to be analyzed by GC, compounds in the samples must be able to get volatilized and additionally possess thermostability.

In case of LC, solubility in the mobile phase is important as well as size, structure and hydrophobicity, presence of functional groups, etc.

Regarding matrix complexity it could be also mentioned that chromatography can be used both for the separation of a compound from the matrix and for quantification and identification. It is important but rarely considered that no residual matter should remain in the samples; especially heavy compounds, which are difficult to evaporate in the GC columns, could significantly influence subsequent analyses. Thus, regular control of retention times and response factors, as well as column cleaning or replacement in due time should not be overlooked. For some samples related to the analysis of biomass even prefractionation could be necessary.

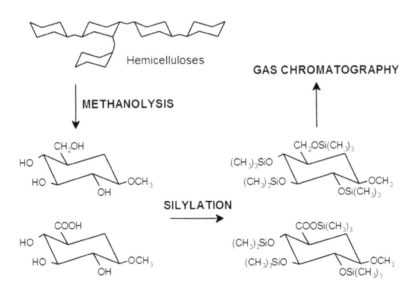

FIGURE 10: Analysis of sugar units in hemicellulose

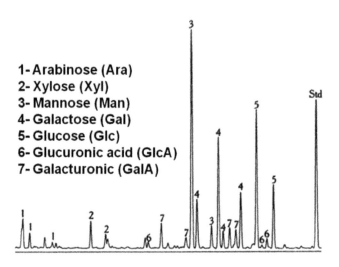

FIGURE 11: Typical gas chromatogram showing the major sugar units released upon methanolysis of a sample (spruce wood) containing hemicelluloses (Std = internal standard, sorbitol)

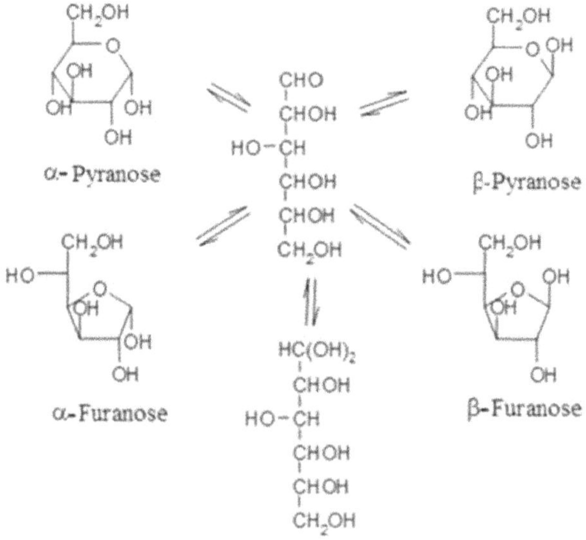

FIGURE 12: Equilibrium of different forms of sugars

3.4 ANALYTICAL EXAMPLES

3.4.1 ANALYSIS OF CARBOHYDRATES AND THEIR TRANSFORMATION PRODUCTS

Let us consider first the analysis of a sample of hemicelluloses dissolved in water. The general analytical strategy is given in Figure 10. An analytical procedure using GC based on acid methanolysis consists of the following steps [23]. Freeze drying of a 2 mL solution of hemicellulose in water with the subsequent addition of 2 M HCl in water-free methanol, is followed by keeping the sample at 100 °C for three hours of neutralization with pyridine, addition of internal standard (sorbitol), evaporation, silylation (hexamethyldisilazane and trimethylchlorosilane), and finally GC analysis. The latter could use, for instance, a split injector (260 °C, split ratio 1:15) with a 30 m/0.32 mm i.d. column coated with dimethyl polysiloxane (DB-1, HP-1), hydrogen or helium as a carrier gas and FID with a following temperature programme: 100–280 °C and ramping 4 °C/min.

An advantage of direct methanolysis of wood samples is that essentially only hemicelluloses are cleaved and very little cellulose. Moreover, contrary to hydrolysis, it allows less degradation of released monosaccharides. Methanolysis can be used also for direct analysis of solid wood and fiber samples. A typical chromatogram is presented in Figure 11, showing several peaks for a particular sugar due to the presence of α & β anomers of pyranoses & furanoses (Figure 12).

Due to the complexity of the product mixture and the analytical procedure correction factors are needed. For instance, cleavage (the methanolysis) could be incomplete for certain glycosidic bonds, such as the Xyl-MeGlcA bond. Some degradation of formed sugars, especially uronic acids may happen and the products can have different detector responses. In order to determine correction factors it is recommended to perform methanolysis, silylation and GC analysis on a sample containing equal amounts of Ara, Xyl, Man, Glc, Gal, GlcA, GalA, etc., and pure hemicelluloses and pectins (if present) and to compare peak areas with the area of the internal standard.

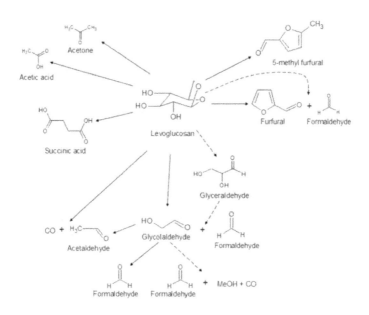

FIGURE 13: Transformations of levoglucosan [24]

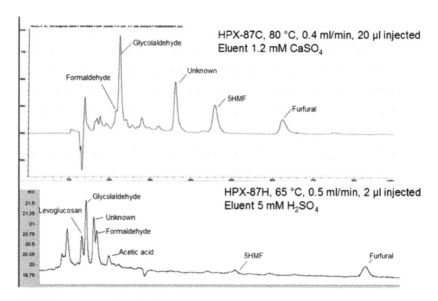

FIGURE 14: Analysis of a levoglucosan transformation mixture by HPLC with two different columns [26]

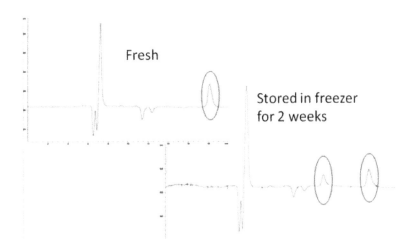

FIGURE 15: HPLC data showing instability of reaction products in levoglucosan transformations [26]

Another example worth considering is the gas-phase catalytic transformation of levoglucosan over zeolites [24,25]. The reaction scheme is given in Figure 13. In [24,25] for HPLC analysis an acid Aminex cation H^+ column with sulfuric acid (0.005 M) as a mobile phase with a flow of 0.5 ml/min at 338 K was used, along with an Aminex HPX-87C column and mobile phase-calcium sulfate (1.2 mM) with a flow rate of 0.4 ml/min at 353 K. A refractive index detector was applied. Figure 14 illustrates that the separation is very much dependent on the analytical conditions.

Stability of the samples is another important issue, which should also be carefully considered, as illustrated in Figure 15. Samples stored in a freezer exhibited another peak, which is certainly a result of transformations happening during storage.

An even more prominent difference in analysis was noticed in the aqueous reforming of sorbitol [27-29]. Comparison of the analysis for different columns is given in Figure 16 demonstrating that for the identification of reaction products tedious and time-consuming analytical work is required.

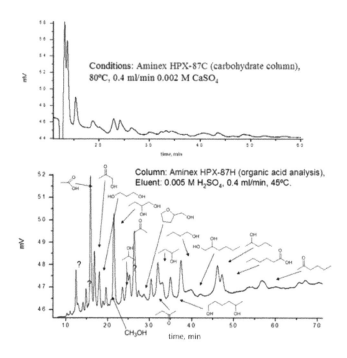

FIGURE 16: HPLC analysis of aqueous phase reforming products [29]

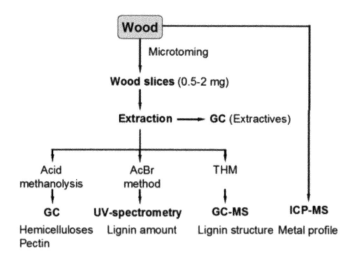

FIGURE 17: A scheme for the microanalysis of wood [30]

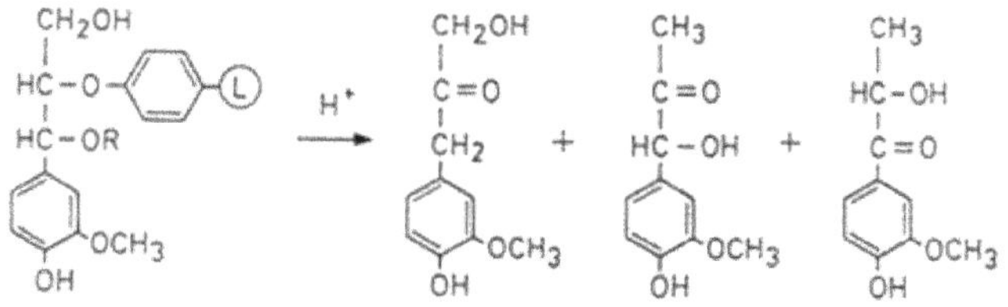

FIGURE 18: Acidolysis of lignin

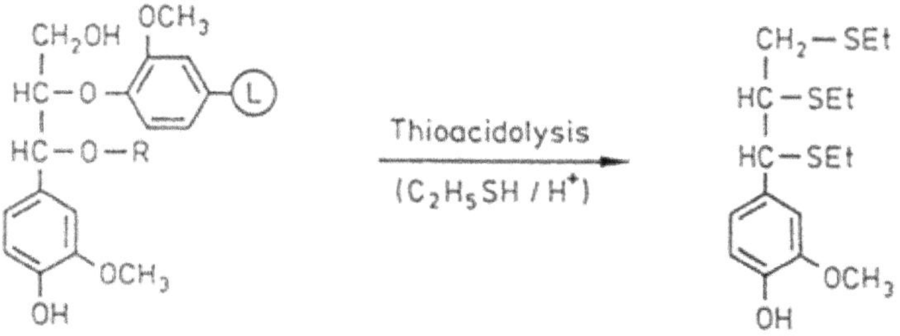

FIGURE 19: Thioacidolysis of lignin

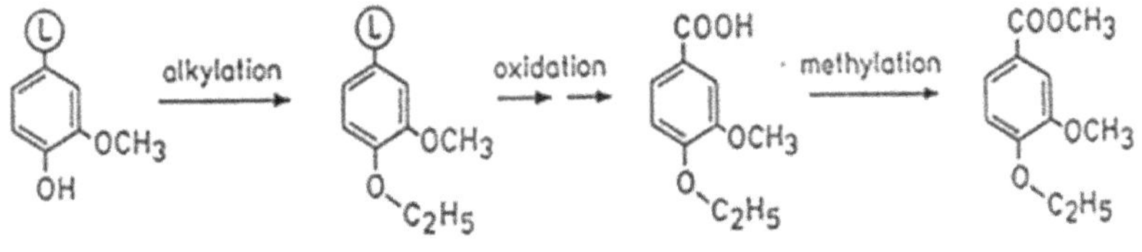

FIGURE 20: Oxidation method for the analysis of terminal units (free phenolic groups) in lignin

3.4.2 ANALYSIS OF LIGNIN

Structural information on lignins could be obtained by wet-chemical and spectroscopic methods using the approach for analysis of wood given in Figure 17. Here, 5 mL 20% AcBr in pure acetic acid is added to ca. 1–10 mg of wood followed by the addition of 0.1 mL perchloric acid (70%) and keeping the mixture for 3 hours at 50 °C, with subsequent neutralization with NaOH, dilution and UV-Vis analysis at 280 nm.

Direct analysis of lignin in wood can be performed by selective / specific degradation followed by GC analysis. Among degradation methods acidolysis, thioacidolysis, permanganate oxidation and pyrolysis can be mentioned. Acidolysis (Figure 18) cleaves predominantly β-O-4-ether bonds by acid hydrolysis and gives many degradation products with a rather low yield of ca. 60%.

Thioacidolysis (Figure 19) gives selective cleavage of β-O-4-ether bonds and results in less complex mixtures than acidolysis and also gives higher yields (> 80%) being able to quantify units with β-O-4-ether bonds and a free hydroxyl. This reaction is performed in dioxane-ethanethiol with boron trifluoride etherate. The degradation products are silylated prior to analysis by GC.

Oxidation by permanganate ($KMnO_4$ – $NaIO_4$ at 82 °C for 6 hours) in acidic solution of ethylated (at pH = 11 and 25 °C with diazoethane for 25 hours) samples with further oxidation by alkaline peroxide for 10 min at 50 °C and final methylation results in samples which can be analyzed by GC (Figure 7.20). This method gives information only on aromatic units with a free hydroxyl, comprising about 10% of lignin in wood and ca. 70% of lignin after kraft cooking.

Another method for the analysis of lignin is pyrolysis combined with GC-MS (Py GC-MS) allowing even a simultaneous determination of lignin and carbohydrates. Py GC-MS can be combined with advanced chemometric methods such as principal component analysis to enable a more complete identification of various lignin fragments. In summary, it can be stated that because of the heterogeneity of lignin there is no universal degradation method giving all desired information on the lignin structure, however, by combination of several methods the structure of lignin can be described fairly well.

Location in the wood	Resin canals (Oleoresin)	Parenchyma cells	Heartwood	Cambium and growth zone	Ascending water Sap
Major compound classes	Resin acids Monoterpenoids Other terpenoids	Fats, fatty acids Steryl esters Sterols	Phenolic substances	Glycosides Sugars, starch proteins	Inorganics
Main function in the tree	Protection	Physiological food reserve, cell membrane comp.	Protection	Biosynthesis Food reserve	Photosynthesis Biosynthesis
Solubility					
Alkanes	+++	+++	0	0	0
DCM	+++	+++	++	0	0
Acetone	+++	+++	+++	++	+
Water	0	0	0	+++	++

FIGURE 21: Classification of wood extractives [31]

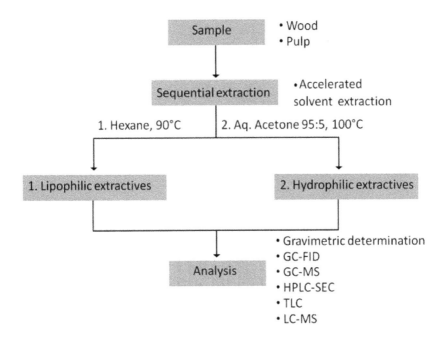

FIGURE 22: Analytical procedures for wood extractives

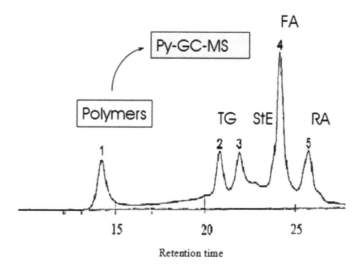

FIGURE 23: Separation of wood extractives with SEC. TG, StE, FA and RA stand for triglycerides, sterols, fatty acids and resin acids respectively [32].

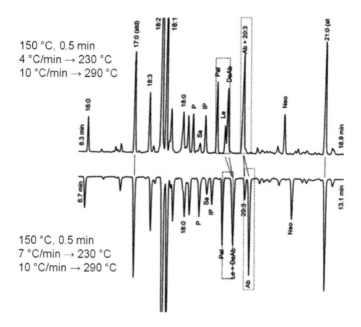

FIGURE 24: GC of fatty acids and resin acids with HP-1, 30 m, 0.32 mm i.d. column with different temperature gradients [31]

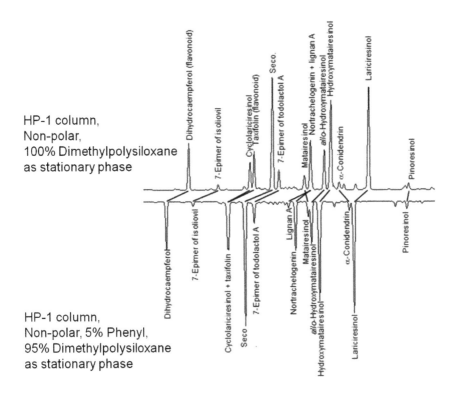

HP-1 column,
Non-polar,
100% Dimethylpolysiloxane
as stationary phase

HP-1 column,
Non-polar, 5% Phenyl,
95% Dimethylpolysiloxane
as stationary phase

FIGURE 25: GC separation of phenolic extractives (flavonoids and lignans) using columns of different polarity [21]

3.4.3 WOOD EXTRACTIVES

Wood contains a wide variety of components that are extractable with various organic solvents or water. Non-polar and semi-polar solvents (hexane, dichloromethane, diethyl ether, MTBE etc.) extract lipophilic oleoresin and fat components, while polar solvents (acetone, ethanol, water, etc.) extract hydrophilic phenolics, sugars, starch and inorganic salts. Acetone and ethanol extract also lipophilic extractives.

A classification of wood extractives is given in Figure 21, while the analytical procedure for extractives is outlined in Figure 22. Group analysis

of fatty acids, resin acids, triglycerides, lignans and sterols can be done using a short column GC (5–7 m/0.53 mm capillary column with 0.15 μm film
thickness), or by HP-SEC (Figure 23) as well as thin layer chromatography.

The analysis of individual compounds can be done by GC on a longer
column (20–30 m/0.20–0.32 mm capillary columns) and reverse phase
HPLC, while the identification of compounds can be performed by GC-
MS, LC-MS, and NMR of isolated substrates. In case of a poor separation
between compounds, the following parameters could be modified: temperature gradient, column polarity, type of derivative used in derivatization. As seen in Figure 24 a better separation, for example, between abietic
acid and tri-unsaturated C_{20} fatty acid, is achieved with somewhat higher
ramping. The HP-1 column usually follows a boiling point order, however,

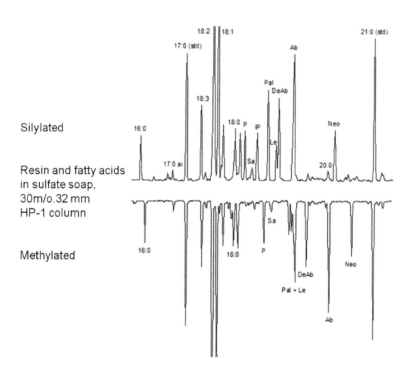

FIGURE 26: Comparison between GC analysis of methylated and silylated fatty and resin
acids [31]

columns with different polarity could also be used (Figure 25) to allow better separation.

As previously mentioned, derivatization of fatty and resin acids is needed for accurate quantitative analysis. Although methylation is a commonly used method, silylation can sometimes afford better separation (Figure 26). In addition, for some GC columns peak-tailing is more severe for methyl esters than for silyl esters.

3.5 FINAL WORDS

This chapter describes some of the contemporary methods for the chemical analysis of biomass-derived chemicals. All available methods could not have been treated in this review, therefore the focus was mainly on chromatographic methods. A more comprehensive overview of analytical methods was published several years ago by one of the authors [31,33]. In the current work, detailed procedures were discussed for only a few cases as the emphasis was laid more on general approaches.

The analytical procedure depends very much on the objective of a particular study as well as the available resources in terms of instruments, time, costs and human skills.

The main hurdles on the way toward the development of a reliable analytical method for a particular application are associated with a lack of time to check methods described in literature, a certain trust in already published procedures, even if they are far from being perfect, as well as a pressure from granting agencies/sponsors to get "real" catalytic results rather than means to develop or check analytical methods. In the latter case there is certainly more glory in developing new methods compared to just checking the old ones.

Finally, we should stress that no single method works perfectly for all kinds of samples. Moreover, dubious methods are sometimes presented in literature, which means that the results are not reliable. It can thus be emphasized once more that improving analytical methods, not only in the particular case of the catalytic transformation of biomass, but in general improves the quality of science.

BIBLIOGRAPHY

1. C. Okkerse, H. Bekkum. From Fossil to Green Green Chemistry 4: 107-114 (1999): 107-114.
2. P.T. Anastas, J.C. Wagner. Green Chemistry Theory and Practice. New York: Oxford University Press, 1998
3. T. Werpy, G. Petersen. Top Value Added Chemicals from Biomass, Vol 1: Results of Screening for Potential Candidates from Sugars and Synthesis Gas. Battelle: US Department of Energy, Energy EfficiencyRenewable Energy,, 2004
4. G.W. Huber, S. Iborra, S. I.. Synthesis of Transportation Fuels from Biomass: Chemistry, Catalysts, and Engineering Chemical Reviews 106: 4044-4098 (2006): 4044-4098.
5. L. Petrus, M.A. Noordermeer. Biomass to Biofuels, a Chemical Perspective Green Chemistry 8: 861-867 (2006): 861-867.
6. G.W. Huber, A. Corma. Synergies between Bio- and Oil Refineries for the Production of Fuels from Biomass Angewandte Chemie International Edition (2007):.
7. S. Lestari, P. Mäki-Arvela, P. MA., Beltramini P., B. P.. Transforming Triglycerides and Fatty Acids to Biofuels ChemSusChem 2: 1109-1119 (2009): 1109-1119.
8. R. Rinaldi, F. Schüth. Design of Solid Catalysts for the Conversion of Biomass Energy & Environmental Science 2: 610-626 (2009): 610-626.
9. D.M. Alonso, J.Q. Bond, J. B., Serrano-Ruiz J.Q.. Catalytic Conversion of Biomass to Biofuels Green Chemistry 12: 1493 (2010): 1493.
10. J. Zakzeski, P.C.A. Bruijnincx, P. B., Jongerius P.C.A.. The Catalytic Valorization of Lignin for the Production of Renewable Chemicals Chemical Reviews 110: 3552-3599 (2010): 3552-3599.
11. E. Adler. Lignin Chemistry - Past, Present and Future Wood Science and Technology 11: 169-218 (1977): 169-218.
12. A. Atutxa, R. Aguado, R. A., Gayubo R., G. R.. Kinetic Description of the Catalytic Pyrolysis of Biomass in a Conical Spouted Bed Reactor Energy Fuels 19: 765-774 (2005): 765-774.
13. P. Mäki-Arvela, B. Holmbom, B. H., Salmi B.. Recent Progress in Synthesis of Fine and Specialty Chemicals from Wood and Other Bio- mass by Heterogeneous Catalytic Processes Catalysis Reviews, Science and Engineering 49: 197 (2007): 197.
14. A. Fukuoka, P.L. Dhepe. Catalytic Conversion of Cellulose into Sugar Alcohols Angewandte Chemie International Edition 45: 5161-5163 (2006): 5161-5163.
15. P.L. Dhepe, A. Fukuoka. Cellulose Conversion under Heterogeneous Catalysis ChemSusChem 1: 969-975 (2008): 969-975.
16. A. Fukuoka, P.L. Dhepe. Sustainable Green Catalysis by Supported Metal Nanoparticles The Chemical Record 9: 224-235 (2009): 224-235.
17. V. Jolle, F. Chambon, F. C., Rataboul F., R. F., F. Rataboul, F. R.. Non-Catalyzed and Pt/Gamma-Al Green Chemistry 11: 2052-2060 (2009): 2052-2060.
18. F. Oersa, B. Holmbom. A Convenient Method for the Determination of Wood Extractives in Papermaking Process Waters and Effluents Journal of Pulp and Paper Science 20: J361-J365 (1994): J361-J365.

19. S. Willför, A. Pranovich, A. P., Tamminen A., T. A., T. Tamminen, T. T., Puls T., P. T., J. Puls, J. P.. Carbohydrate Analysis of Plant Materials with Uronic Acid-Containing Polysaccharides - A Comparison between Different Hydrolysis and Subsequent Chromatographic Analytical Techniques Industrial Crops and Products 29: 571-580 (2009): 571-580.

20. P. Tolvanen, P. Mäki-Arvela, P. MA., Kumar P., K. P., N. Kumar, N. K., Eraenen N., E. N.. Thermal and Catalytic Oligomerisation of Fatty Acids Applied Catalysis A: General 330: 1-11 (2007): 1-11.

21. S.M. Willför, A.I. Smeds, A. S.. Chromatographic Analysis of Lignans Journal of Chromatography A 1122: 64-77 (2006): 64-77.

22. I.V. Deliy, N.V. Maksimchuk, N. M., Psaro N.V., P. N., R. Psaro, R. P., Ravasio R., R. R.. Kinetic Peculiarities of cis/trans Methyl Oleate Formation during Hydrogenation of Methyl Linoleate over Pd/MgO Applied Catalysis A: General 279: 99-107 (2005): 99-107.

23. A. Sundberg, K. Sundberg, K. S., Lillandt K.. Determination of Hemicelluloses and Pectin in Wood and Pulp Fibres by Acid Methanolysis and Gas Chromatography Nordic Pulp & Paper Research Journal 226: 216-219 (1996): 216-219.

24. M. Käldström, N. Kumar, N. K., Heikkilä N., H. N., T. Heikkilä. Formation of Furfural in Catalytic Transformation of Levoglucosan over Mesoporous Materials ChemCatChem 2: 539-546 (2010): 539-546.

25. M. Käldström, N. Kumar, N. K., Salmi N.. Levoglucosan Transformation over Aluminosilicates Cellulose Chemistry and Technology 44: 203-209 (2010): 203-209.

26. M. Käldström, D.Yu. Murzin. unpublished data. .

27. A.V. Kirilin, A.V. Tokarev, A. T., Murzina A.V., M. A., E.V. Murzina. Reaction Products and Intermediates and their Transformation in the Aqueous Phase Reforming of Sorbitol ChemSusChem 3: 708-718 (2010): 708-718.

28. A.V. Tokarev, A.V. Kirilin, A. K.. unpublished data. .

29. A.V. Tokarev, A.V. Kirilin, A. K., Murzina A.V., M. A., E.V. Murzina, E. M.. The Role of Bioethanol in Aqueous Phase Reforming to Sustainable Hydrogen International Journal of Hydrogen Energy 35: 12642-12649 (2010): 12642-12649.

30. A. Pranovich, J. Konn, J. K.. Methodology for Chemical Microanalysis of Wood. In: Proceedings 9th European Workshop on Lignocellulosics and Pulp . 2006. 436-439.

31. B. Holmbom. Extractives. In: Analytical Methods on Wood Chemistry, Pulping and Papermaking Berlin: Springer Verlag, 1998. 125-148.

32. B. Holmbom. unpublished data. .

33. B. Holmbom, P. Stenius. Papermaking Science and Technology. In: Forest Products Chemistry , 2000. 105-169.

PART II

REACTION ROUTES

CHAPTER 4

Catalytic Routes for the Conversion of Biomass Into Liquid Hydrocarbon Transportation Fuels

JUAN CARLOS SERRANO-RUIZ AND JAMES A. DUMESIC

4.1 INTRODUCTION

Society has reached high levels of development during the last century. This progress, however, has been achieved at the expense of extensive consumption of natural resources, such as petroleum, natural gas and coal. These fossil fuel resources took millions of years to be formed, and they are currently being consumed at a rate that is orders of magnitude higher than their natural regeneration cycle, making them non-renewable sources of energy. The most recent data available for world energy consumption indicate that society still remains highly dependent on fossil fuels at the present time. For example in 2008, fossil fuels supplied 85% of the total energy consumed in the US, [1] and almost 80% of the energy produced in the European Union. [2] These fossil fuel resources are used to provide energy for various sectors of society (i.e., residential, commercial, indus-

Catalytic Routes for the Conversion of Biomass into Liquid Hydrocarbon Transportation Fuels. © *Serrano-Ruiz JC and Dumesic JA.* Energy & Environmental Science *4,83 (2011), DOI: 10.1039/ c0ee00436g. Reproduced from* Energy & Environmental Science *with permission from The Royal Society of Chemistry.*

trial, transportation and electrical power), among which the transportation sector is the largest and fastest growing energy sector, responsible for almost one third of the total energy consumed in the world. Moreover, a large fraction of the energy for the transportation sector (96%) is currently derived from petroleum. [3]

Three important issues are associated with the large-scale utilization of fossil fuels: availability, global warming and uneven geographic distribution of reserves. Fossil fuels are finite and, as indicated above, their current consumption rate is higher than their corresponding regeneration rate, leading inevitably to depletion. Projections for the near future indicate that world energy consumption will increase by 35% over the next 20 years to meet the growing demand of industrialized countries and the rapid development of emerging economies, [4] and world demand for petroleum will raise by 30%, reaching 111 millions of barrels per day in 2035. [5] Taking into account these forecasts and current data of proven reserves, it has been estimated that oil, natural gas and coal will be depleted within the next 40, 60 and 120 years, respectively. [6] In the case of petroleum, many researchers predict a more dramatic situation and estimate that the global production of oil will reach a maximum in the year 2020 and decay thereafter. [7]

Global warming is, possibly, the most dramatic and known collateral effect produced by the massive utilization of fossil fuels. [8] Fossil fuels are transformed into energy by means of combustion reactions, leading to net emissions of CO_2, a strong greenhouse gas, into the atmosphere. Accordingly, the extraction of fossil fuels for energy production has allowed a large part of the carbon stored in the earth for millions of years to be released in just a few decades.

Fossil fuels reserves are not equally distributed around the world. The Middle-East countries control the 60% of the oil reserves and the 41% of natural gas supplies, and only three countries (US, China and Russia) account for 60% of the world recoverable coal reserves. [4] This situation can lead to economic instabilities, requires the transportation of fossil fuel resources over long distances, and can cause political and security problems worldwide.

The issues outlined above, inherently associated with fossil fuels, suggest that society requires new sources of energy to ensure progress and protect the environment for future generations. These new sources of energy should: (i) have the potential to effectively replace fossil fuels in the current energy production system and (ii) be renewable, well distributed around the world, and not contribute to the accumulation of greenhouse gases into the atmosphere. In this respect, natural resources such as solar energy, wind, hydroelectric power, geothermal activity, and biomass meet these requirements. Unlike fossil fuels, they are abundant and allow the development of zero-carbon or carbonneutral technologies, thus contributing to mitigation of global warming effects. Substitution of fossil fuel-based technologies for those derived from renewable sources is currently spurred by various governments, [9,10] and it will be done progressively and selectively. Thus, while solar, wind, hydroelectric, and geothermal have been proposed as excellent alternatives to coal and natural gas for heat and electricity production in stationary power applications, [11,12] biomass is the only sustainable source of organic carbon currently available on earth, [13] and it is considered to be an ideal substitute for petroleum in the production of fuels, chemicals and carbon-based materials. [14,15] However, when designing strategies for potential replacement of crude oil by biomass, it is important to note that the petrochemical industry currently consumes three quarters of the crude oil to cover the demand for liquid hydrocarbon fuels of the transportation sector, whereas only a small fraction of the petroleum is utilized in the synthesis of industrial chemicals and other derivatives. [16] Consequently, an effective implementation of biomass in the current energy system will necessarily involve the development of new technologies for the large-scale production of biofuels.

At the present time, two biomass-derived fuels (so-called first generation of biofuels) have been successfully implemented in the transportation sector: biodiesel (a mixture of long-chain alkyl esters produced by transesterification of vegetable oils with methanol) and ethanol (produced by bacterial fermentation of corn and sugar cane-derived sugars). The penetration of these liquid biofuels in the transportation sector is still very weak, and in 2005 they represented only 2% of the total transportation energy. [3] However, the important environmental and economic benefits derived from their large-scale utilization will stimulate society to progressively increase reliance on biofuels. Thus, according to projections by the International Energy Agency, the world biofuel production will increase from the current level of 1.9 million of barrels per day (mbd) in 2010 to 5.9 mbd by 2030, which represents 6.3% of the world conventional fuels production. [4] Unlike petroleum-based fuels, liquid biofuels are considered carbon neutral since CO_2 produced during fuel combustion is consumed by subsequent biomass regrowth. [17] Furthermore, recent studies indicate that the use of liquid biofuels produced domestically would strengthen economies by reducing the dependence of foreign oil and by creating new well-paid jobs in different sectors such as agricultural, forest management and oil industries. [18]

4.2 LIGNOCELLULOSIC LIQUID HYDROCARBON FUELS: ALTERNATIVE TO ETHANOL

The current biofuel market is largely dominated by ethanol, which accounts for 90% of world biofuel production. [19] Indeed, the rate of ethanol production around the world is increasing rapidly, from 13 billion gallons in 2007 to the current level of almost 20 billion gallons in 2009, [20] with the US (55%, corn-derived) and Brazil (33%, sugar cane-derived) being the main producers. [21] The ethanol industry has benefited from a mature and simple technology in which selected microorganisms (e.g., yeast, bacteria, and mold) transform aqueous sugars to the desired final ethanol product. Ethanol is added to gasoline to increase the octane number of the mixture and, with it, improve the combustion characteristics of the fuel. The oxygen in ethanol allows low-temperature combustion with the subsequent reduction of pollutants such as CO and NO_x. [19] Additionally, apart from CO_2 emission savings, blending gasoline with ethanol helps to reduce SO_x emissions to the atmosphere, because ethanol contains a negligible amount of sulfur compared to petroleum. [18]

A key aspect that is responsible for expansion of the ethanol industry in recent years is the compatibility of ethanol with the existing infrastructure for gasoline. Thus, ethanol blended with conventional gasoline is currently used in many countries as a renewable fuel in existing spark-ignition engines. This compatibility, however, is not complete and the use of ethanol is presently limited to low-concentration blends (5–10% by volume), namely E5–E10. Ethanol-enriched mixtures such as E85 require cars with specially designed engines, designated as flexible-fuel vehicles (FFVs), which are commonly used only in a few countries like Brazil and Sweden. E85 mixtures are not tolerated by conventional vehicles, because ethanol, especially in high-concentration blends, can cause corrosion of some metallic components in tanks and deterioration of rubbers and plastics used in internal combustion engines. [21] This constraint in ethanol blending represents the main issue of the growing ethanol industry. As outlined in Fig. 1, projections indicate that the US ethanol industry will approach the blending wall (i.e., the point at which blending 10% of ethanol in each gallon of gasoline will not be able to accommodate the rate of ethanol production) in 2010. [22] Furthermore, experts predict that the number of E85 fuelling stations and flexi-fuel vehicles will not grow sufficiently fast to accommodate the growing volumes of ethanol produced in the US. [23] A potential solution to overcome the blending wall is to raise the amount the ethanol allowed in gasoline to beyond 10% by commercializing intermediate ethanol blends (i.e., E15– E20) (Fig. 1). However, there are serious issues in using blends with higher ethanol concentrations. In countries like the US, the utilization of E15–E20 blends in regular vehicles is still not authorized, since the effect of these mixtures on pollutant emissions, driving performance and materials compatibility (e.g., tanks, pipelines, dispensers) is not fully understood, [23] and current European standards allow only for E5 blends. [18]

Apart from the aforementioned blending issues, ethanol presents another important limitation as a transportation fuel. Ethanol contains less energy per volume (i.e., energy density) than conventional gasoline, which ultimately reduces the fuel mileage of the vehicles. In this sense, it has been estimated that cars running on ethanol rich mixtures like E85 operate with 30% lower fuel mileage. [24] This fact, along with the small price differential between E85 and regular gasoline, has discouraged drivers to purchase E85 cars or fuel so far. [22]

Conventional transportation fuels are composed of liquid hydrocarbons with different molecular weights (e.g., C5–C12 for gasoline, C9–C16 for jet fuel, and C10–C20 for diesel applications) and chemical structures (e.g., branched for gasoline, linear for diesel). The entire transportation infrastructure (including engines, fueling stations, distribution networks, and storage tanks) has been developed to take advantage of the excellent properties of these compounds as fuels. Thus, the special composition of hydrocarbons fuels, based only on carbon and hydrogen, provides them with high energy-density and stability (allowing efficient storage at ambient conditions) and superior combustion characteristics, properties highly desired for transportation liquids. Thus, instead of using biomass to produce oxygenated fuels (such as ethanol) with new compositions, an attractive alternative would be to utilize biomass to generate liquid fuels chemically similar to those being used today derived from oil. [17,25] These new fuels would be denoted as green gasoline, green diesel and green jet fuel, and they would be essentially the same as those currently used in the transportation fleet, except that they would be synthesized from biomass instead of petroleum. When compared with ethanol, the production of hydrocarbon fuels from biomass has important advantages. The main benefit would include full compatibility with the existing energy system. Since green hydrocarbon fuels would be essentially the same as those currently obtained from petroleum, it would not be necessary to modify engines, pumps or distribution networks to accommodate the new renewable liquids in the transportation sector.

Unlike ethanol, biomass-based hydrocarbons fuels are energy equivalent to fuels derived from petroleum. The heating value (i.e., the heat released when a known quantity of fuel is burned under specific conditions) of ethanol is only two-thirds that of gasoline, which, as indicated above, penalizes the fuel mileage of the vehicles running on gasoline–ethanol mixtures. The use of renewable hydrocarbon fuels would additionally help to meet the increased standards of fuel economy imposed by governments to the automobile industry. In the case of the US, these standards establish a mandatory increase in average fuel economy from the current 25 miles per gallon (mpg) to 35 mpg by 2022. [26]

The addition of oxygenated components to conventional fuels increases the water solubility of the mixture. This increase is particularly marked in the case of gasoline–ethanol blends, since pure ethanol is highly hygroscopic and completely miscible in water. Thus, adding 10% of ethanol to regular gasoline raises the water solubility of the blend more than 30 times (from 150 ppm v/v of regular gasoline to 5000 ppm for E10). [27] Once the water contamination reaches the saturation level, additional water separates from the mixture, removing the ethanol from gasoline and leading to phase separation. In fact, when phase separation occurs in the storage tank, the ethanol–water layer may combust in the engine at higher temperatures causing damage to it. [27] The water tolerance of a gasoline–ethanol blend (i.e., fraction of water that the mixture can contain without phase separation) decreases with temperature and increases with ethanol content (Fig. 2). Consequently, phase separation is an important concern in countries with cooler climates and when low-concentration blends such as E5 are used. Water can be absorbed by the ethanol–gasoline mixture from the atmosphere (in the form of moisture), from the air trapped in the tank (by condensation of water when temperature decreases), or even from the ethanol itself which typically carries traces of water when delivered from the biorefinery. In this respect, many countries have regulated the maximum amount of water allowed in fuel-grade ethanol to the level of 1% (v/v), to avoid phase separation issues. [28] Ethanol affinity for water has important implications for distribution logistics as well. Pipelines, considered to be the least expensive means of safely transporting bulk fuel shipments, [23] are not suited to transport ethanol or gasoline–ethanol blends on a commercial scale, because apart from corrosion issues, ethanol can pick up water in the pipeline with the potential result of phase-separation. Consequently, ethanol has to be distributed by other fossil fuel-consuming transportation modes such as rail, truck and barge. The hydrophobic character of biomass-derived hydrocarbons eliminates these problems, since these molecules are immiscible in water. Additionally, the ability of liquid hydrocarbons to self-separate from water, as represented in Fig. 3, is highly beneficial in that it eliminates the need for expensive and energy-consuming distillation steps required in the ethanol purification process. In particular, ethanol is initially obtained in form of a dilute aqueous solution (5–12% v/v), which is subsequently concentrated to 96–99% by distillation. It is estimated that this intense water removal step is responsible for 35–40% of the total energy required for ethanol

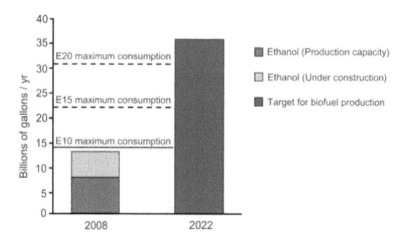

FIGURE 1: Volumes of ethanol absorbed by several blends in the US. Target for biofuels production refers to the level of biofuels production mandated by the Energy Independence Security Act of 2007 (EISA 2007). [26] Maximum consumption for a determined blend refers to the ethanol consumed if all the gasoline used in the country is blended with ethanol in the amount indicated. Source: Biomass Research and Development Board. [23]

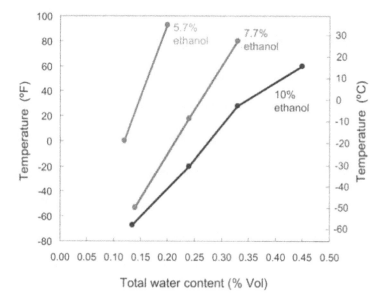

FIGURE 2: Water tolerance of some gasoline–ethanol blends as a function of temperature. Adapted from ref. 27.

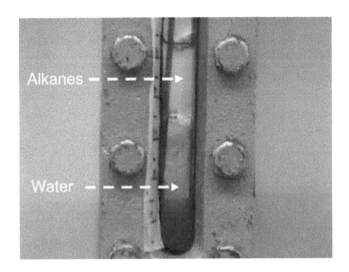

FIGURE 3: Picture showing spontaneous separation between aqueous phase and liquid alkanes produced by processing of biomass-derived molecules in a four-phase flow reactor. Source: Dumesic's Research Group web site, http://jamesadumesic.che.wisc.edu/index2.htm.

production, [29] and this energy is typically supplied by combustion of fossil fuels such as natural gas.

Any technology envisaged to convert biomass feedstocks into liquid fuels must address one important limitation of this resource: the low energy-density of biomass compared to fossil fuels. Although the energy-density of biomass varies considerably depending on the source, an average value for biomass (15–20 MJ kg^{-1}) is well below that of crude oil (42 MJ kg^{-1}). [30] Large amounts of biomass will thus be required to produce liquid fuels, leading to high costs for transporting the biomass source to the processing location. [31] Furthermore, if biomass transportation involves utilization of fossil fuels, then the overall CO_2 emission savings of the bioprocess would be penalized. Consequently, for biomass conversion technologies to be cost-competitive and truly carbon-neutral, it is necessary to develop efficient processing units at small scale that can be distributed close to the biomass source. [17] Even though ethanol plants achieve a significant size reduction compared to petrochemical refineries, the mild conditions employed and the low levels of ethanol achieved for bacterial fermentation (e.g., 30–50 °C, ethanol concentrations lower than 15% v/v) significantly limit the reaction rates and require reactors with a size large enough to make the process economically feasible. As will be described in following sections, biomass-based hydrocarbon fuels, in contrast, can be produced at high temperatures and using concentrated water solutions, [14] which allow for faster conversions in smaller reactors.

The utilization of edible biomass (such as corn or cane sugar) for the large-scale production of fuels can produce competition with food for land use. The so-called food-versus-fuel debate has arisen in many countries as a response to the sharp increase in food prices during 2007 and 2008. Although it has been pointed out that this rise in price was the result of several linked worldwide events, [22] some authors indicate that the increased demand for corn to produce ethanol had a direct impact on food prices, especially in food-insecure areas of the world where food is based on grain consumption. [32] These issues have driven researchers around the world to develop technologies to process non-edible biomass (e.g., lignocellulosic biomass), thereby permitting sustainable production of a new generation of biofuels (so-called second generation of fuels), without affecting food supplies. Lignocellulosic biomass has two important advantages over edible biomass feedstocks: it is more abundant and can be grown faster and with lower costs. [33] In this respect, it is estimated that the US could sustainably produce more than 1 billion tons of non-edible biomass per year by 2050 with relatively modest changes in land use and agricultural and forestry practices. Once converted into biofuels, these lignocellulosic resources would have the potential to displace more than one third of the petroleum currently consumed by the transportation sector.34 Lignocellulosic feedstocks ($3 per GJ) are slightly less expensive than edible biomass (5$ per GJ), and potentially more economical than crude oil (10–15 $ per GJ) and vegetable oils (18–20 $ per GJ); however, due to its recalcitrant nature, lignocellulose is more difficult to convert and, consequently, processing costs increase for this resource. [35] Thus, according to recent analyses, the cost of producing lignocellulosic ethanol would be almost double that of corn-derived ethanol. [19] This fact represents the main limitation of lignocellulose as a renewable resource, and the lack of cost-competitive technologies for the generation of liquid fuels from non-edible sources has been identified by experts as the key bottleneck for the large-scale implementation of lignocellulose- derived biofuel industry. [36]

Recalcitrance of lignocellulosic biomass can be explained in terms of its chemical structure, comprised of three major units: cellulose, hemi-cellulose and lignin. [13,37,38] Cellulose (40–50%) is a high molecular weight polymer of glucose units connected linearly via b-1,4-glycoside linkages. This arrangement allows for extensive hydrogen bonding be-tween cellulose chains, which confers this material with rigid crystallinity and, thus, high resistance to deconstruction. [39] Cellulose bundles are additionally attached together by hemicellulose (15–20%), an amorphous (and consequently more readily deconstructed) polymer of five different C5 and C6 sugars. Cellulose and hemicellulose, the carbohydrate frac-tion of lignocellulose, are protected by a surrounding three-dimensional polymer of propyl-phenol called lignin (15–25%), which provides extra rigidity to the lignocellulose structure. To overcome lignocellulose recal-citrance, a variety of physical and chemical methods have been developed, and a comprehensive description of such technologies can be found else-where. [40–42] The approach most commonly used involves pretreatment of lignocellulose (with the aim of breaking/ weakening the lignin protec-tion and increasing the susceptibility of crystalline cellulose to degrada-tion), followed by hydrolysis to depolymerize hemicellulose and cellulose and, thus, isolate the sugars from the lignin fraction.

4.3 ROUTES FOR THE PRODUCTION OF BIOFUELS FROM BIOMASS

The main pathways for the production of liquid transportation fuels from biomass are shown in Fig. 4. As indicated in previous sections, food crops such as corn grain or sugar cane can be converted into ethanol by fer-mentation processes. Alternatively, second generation ethanol can be pro-duced from lignocellulosic sources by means of pretreatment–hydrolysis and subsequent fermentation of soluble sugars. Butanol, with energy den-sity and polarity similar to gasoline, can be also produced by this route, representing an interesting alternative to overcome many of the technical shortcomings of ethanol as a fuel. [43,44] Interestingly, the microorganisms utilized for fermentation can be engineered to convert sugars to liquid alkanes instead of alcohols. [45] This new technology could achieve improvements over classical fermentation approaches, because hydrocarbons separate spontaneously from the aqueous phase, thereby avoiding poisoning of microbes by the accumulated products and facilitating separation/ collection of alkanes from the reaction medium.

Vegetable oils, obtained from food sources such as soybeans, palm or sunflower, can serve as feedstocks for the production of first-generation biodiesel through transesterification processes. Since vegetable oils are expensive and compete with food sources, the challenge of the biodiesel industry is to find non-edible sources of lipids. Algae crops are receiving interest in this respect, [46] although the high cost associated with feedstock production is an important barrier, and related technologies are presently at an early stage of development. Green diesel can be produced from plant oils and animal fats by means of deoxygenation reactions under hydrogen pressure in hydrotreating processes. [47,48] This recent technology has potential in that it can be carried out in existing petroleum refinery infrastructure. [49]

Representative examples of non-food lignocellulosic feedstocks such as forest wastes, agricultural residues like corn stover, or municipal paper wastes are shown in Fig. 4. Apart from their intrinsic recalcitrance, these feedstocks are characterized by a high degree of chemical and structural complexity, and, consequently, technologies for the conversion of these resources into liquid hydrocarbon fuels typically involve a combination of different processes. The methodology most commonly used to overcome lignocellulose complexity involves the transformation of non-edible feedstocks into simpler fractions that are subsequently more easily converted into a variety of useful products. This approach, similar to that used in conventional petroleum refineries, would allow the simultaneous production of fuels, power, and chemicals from lignocellulose in an integrated facility denoted as a biorefinery. [50,51] Current technologies for converting lignocellulose to liquid hydrocarbon transportation fuels involve three major routes: gasification, pyrolysis and pretreatment–hydrolysis (Fig. 4). By means of these primary routes, lignocellulose is converted into gaseous and liquid fractions that are subsequently upgraded to liquid hydrocarbon fuels. Thus, gasification converts solid biomass to synthesis gas (syngas), a valuable mixture of CO and H_2 which serves as a precursor of liquid hydrocarbon fuels by Fischer– Tropsch (F–T) reactions. This pathway is commonly known as biomass to liquids (BTL). Pyrolysis allows transformation of lignocellulosic biomass into a liquid fraction known as bio-oil that can be subsequently upgraded to hydrocarbon fuels by a variety of catalytic processes. The third route involves pretreatment–hydrolysis steps to yield aqueous solutions of C5 and C6 sugars derived from lignocellulose. While gasification and pyrolysis are pure thermal routes in which lignocellulose is decomposed with temperature under controlled atmosphere, aqueous-phase processing, in contrast, involves a series of catalytic reactions to selectively convert sugars and important platform chemicals derived from them into targeted liquid hydrocarbon fuels with

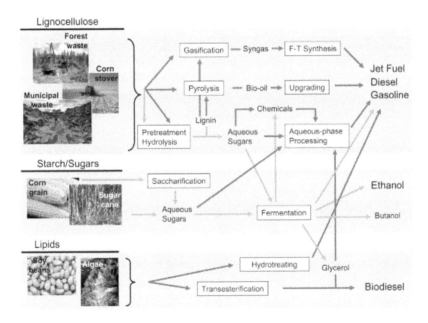

FIGURE 4: Routes for the conversion of biomass into liquid fuels. Red arrows refer to thermal routes, green arrows refer to biological routes, and blue arrows refer to catalytic routes. Adapted from ref. 25.

molecular weights and structures appropriate for gasoline, diesel and jet fuel applications.

4.4 PRODUCTION OF LIQUID HYDROCARBON TRANSPORTATION FUELS FROM LIGNOCELLULOSE

4.4.1 BIOMASS TO LIQUIDS (BTL)

Biomass to liquids can be described as a renewable version of fossil fuel-based technologies like coal to liquids (CTL) and gas to liquids (GTL), involving the integration of two different processes: biomass gasification to syngas (H_2/CO) and F–T synthesis. Even though both technologies are well known and relatively mature, integration remains a challenge in BTL, because the utilization of lignocellulosic biomass as a feedstock (in substitution for classical carbon sources such as coal and natural gas) introduces new difficulties in the overall process. Biomass gasification is achieved by treatment at high temperatures (e.g., 1100–1500 K) under a well-controlled oxidizing atmosphere (e.g., air, steam, oxygen). Control over the composition of the outlet gaseous stream is difficult and depends on a variety of factors including the oxidizing agent, biomass particle size and gasifier design. [52] In this respect, research indicates that utilization of pure oxygen atmosphere, small particle sizes (lower than 1 mm diameter), and a combination of high temperatures, high pressures and low residence times favors the production of syngas versus producer gas (a mixture of CO, H_2, CO_2, CH_4, and N_2 used for heat and electricity production). [53–55]

The direct integration of biomass gasification and F–T synthesis requires an intermediate gas-cleaning system, because the gaseous stream delivered from the gasifier typically contains a number of contaminants that need to be removed before the F–T unit, which is highly sensitive to impurities. Thus, tars (condensable high molecular weight hydrocarbons produced by incomplete biomass gasification), volatile species such as NH_3, HCl, and sulfur compounds (produced by gasification of lignocellulose impurity components), fine particles, and ashes typically accompany CO and H_2 in the outlet gaseous stream. The high number of contaminants, along with the strict cleaning standards imposed by the F–T unit,

[56] require the use of multiple steps and advanced technologies [52] that contribute significantly to the complexity and cost of the BTL plant. Additionally, because biomass contains higher amounts of oxygen compared to coal, the syngas delivered from lignocellulosic sources is typically enriched in CO (H_2/CO ¼ 0.5), and F–T synthesis requires syngas with a H_2/CO ratio closer to 2.57, [58]. By providing sufficient water co-feeding, the H_2/CO ratio can be adjusted by means of an intermediate water gas-shift (WGS, CO + H_2O / CO_2 + H_2) reactor situated between the gasifier and the F–T unit.

The F–T reactor, the last unit of the BTL plant, achieves conversion of syngas to a distribution of alkanes over Co-, Fe-, or Ru-based catalysts in a well-developed industrial process. [58] However, the hydrocarbons produced by the direct route range from C_1 to C_{50}, and neither gasoline nor diesel fuels can be produced selectively without generating a large amount of undesired products. Indirect approaches involve initial production of heavy hydrocarbons (waxes), followed by controlled cracking of the heavy compounds to diesel and gasoline components to overcome this limitation. [59]

The cost of producing the biofuel, negatively affected by the complexity of the process, is the main factor limiting the commercialization of BTL technologies. Application of economies- of-scale allows for improvements in the economics of the process at the expense of having large centralized facilities that, as indicated in a previous section, lead to higher costs for transporting the low energy density biomass. BTL profit margins can be increased by co-producing, along with liquid hydrocarbon fuels, higher-value chemicals such as methanol [60] and hydrogen [61,62] from lignocellulose-derived syngas. Another positive aspect of BTL is its versatility. Thus, since any source of lignocellulose can be potentially gasified, BTL technologies are not constrained to a particular biomass feedstock or fraction.

4.4.2 PYROLYSIS INTEGRATED WITH UPGRADING PROCESSES

Lignocellulosic biomass can be treated under inert atmosphere at temperatures of 648–800 K in a process called pyrolysis. At these conditions, solid biomass undergoes a number of processes including depolymerization, dehydration and C–C bond breaking reactions which lead to the formation of reactive vapor species. [35] Upon subsequent cooling, the vapor products condense generating a dark viscous liquid referred to as bio-oil. This bio-oil is a complex mixture of more than 400 highly oxygenated compounds, including acids, alcohols, aldehydes, esters, ketones and aromatic species, along with some remnants of polymeric carbohydrates and lignin fragments. [63,64] Consequently, once separated into their components, bio-oil could serve as a source of chemicals. The final composition of the bio-oil depends on a large number of factors (e.g., feedstock type, reaction conditions, alkali content of the feedstock, storage conditions). Biooils typically contain about 25 wt% water (derived from the initial water of the feedstock and from the pyrolysis process itself), and retain up to 70% of the energy stored in the biomass feedstock, [52] thereby allowing for concentration of the energy of biomass in a dense liquid that is more easily transportable. The main advantage of pyrolysis over BTL is its simplicity, because it requires only a single reactor and low capital investments, thereby allowing the development of cost-effective processing units on small scale. Thus, small portable pyrolysis reactors (i.e., 50–100 tons of biomass per day) are currently commercially developed to produce liquid biofuels close to the biomass location in countries like the US, Canada and the Netherlands. [65,66]

Even though bio-oils can be used directly in simple boilers and turbines for heat and electricity production, their utilization as transportation fuels has multiple shortcomings. The high oxygen content of bio-oils negatively affects the energy density (16–19 MJ kg^{-1} versus 46 MJ kg^{-1} of regular gasoline), and it leads to low volatility and poor stability properties of the bio-oil liquid. Furthermore, the high corrosiveness (pH ~ 2.5) and viscosity of bio-oils discourage their utilization in internal combustion engines. Since the pyrolysis process does not involve a deep chemical transformation in the feedstock, extensive oxygen removal is required for bio-oils to have hydrocarbon-like properties (e.g., high energy density, high volatility and high thermal stability), and several routes are available in this respect (Fig. 5).

Hydrodeoxygenation (i.e., treatment of the bio-oil at moderate temperatures and high hydrogen pressures, HDO) is probably the most common method to achieve oxygen removal from biooils. [67,68] By means of this technology, bio-oil components are completely hydrogenated and oxygen is removed in the form of water, which appears in the reactor as a separate phase from the hydrocarbon layer. Hydrodeoxygenation is typically carried out over sulfided CoMo and NiMo based catalysts [68] (used in the petrochemical industry to achieve sulfur and nitrogen removal from crude oil). Precious metals such as Pt and Ru [69,70] show higher hydrogenation activities at the expense of low tolerance to sulfur impurities (typically present in bio-oils). The large amount of hydrogen required for bio-oil deoxygenation represents the main drawback of this technology, [71] and strategies based on steam-reforming of the water-soluble fraction of bio-oils, [72] along with aqueous-phase reforming of biomass-derived sugars, [73,74] have been studied to avoid the need to supply hydrogen from external fossil fuel sources. Bio-oils typically contain significant amounts of lignin-derived phenols which, once transformed into aromatic hydrocarbons, are valuable gasoline components. [75] One of the challenges of the hydrodeoxygenation process is to achieve complete hydrogenation of aliphatic compounds while avoiding unnecessary hydrogen consumption in the reduction of the valuable aromatic hydrocarbons. However, this control over the extent of the hydrogenation process is difficult at the elevated hydrogen pressures required for hydrodeoxygenation (e.g., 100–200 bars). In addition, high pressures lead to increases in operational costs of the process.

Bio-oil deoxygenation can be alternatively carried out at milder conditions (e.g., 623–773 K, atmospheric pressure) and without external hydrogen by processing the bio-liquid over acidic zeolites, in a route that resembles the catalytic cracking approach used in petroleum refining. [76–78] At these conditions, bio-oil components undergo a number of reactions involving dehydration, cracking and aromatization, and oxygen is removed in the form of CO, CO_2 and water (Fig. 5). As a result, bio-oil is converted into a mixture of aliphatic and aromatic hydrocarbons, although a large fraction of the organic carbon reacts to form solid carbonaceous deposits denoted as coke. Thus, hydrocarbon yields are relatively modest and regeneration cycles under air (to burn off the coke) are frequent. Irreversible deactivation, caused by partial de-alumination of zeolite structures at the water contents typically found in bio-oils, is another drawback of this technology, and research is needed on new acidic catalytic materials with better resistance to water. [16] On the other hand, the conditions of pressure and temperature at which zeolite upgrading is carried out are similar to those used in pyrolysis, thereby allowing the integration of these two processes in a single reactor, as recently demonstrated by Huber et al. [79]

A third route that could help to reduce oxygen content in biooils while leaving the bio-liquid more amenable for subsequent downstream processes is catalytic ketonic decarboxylation or ketonization [80] (Fig. 5). By means of this reaction, 2 molecules of carboxylic acids are condensed into a larger ketone (2n – 1 carbon atoms) with the release of stoichiometric amounts of CO_2 and water. This reaction is typically catalyzed by inorganic oxides such as CeO_2, TiO_2, Al_2O_3 and ZrO_2 at moderate temperatures (573–698 K) and atmospheric pressure. [81–84] Interestingly, ketonization achieves oxygen removal (in the form of water and CO_2) while consuming carboxylic acids, the latter of which represent an important fraction of bio-oils (up to 30 wt%). [85] Moreover, these acids are hydrogen-consuming compounds, and are responsible for unwanted properties of the bio-liquids such as corrosiveness and chemical instability. Consequently, as represented in Fig. 5, a pretreatment of the biooil over a ketonization bed would simultaneously reduce oxygen content and acidity, thereby reducing hydrogen consumption and leaving bio-oil more amenable for subsequent hydrodeoxygenation processing. Even though ketonization has not been used to process real lignocellulosic bio-oils so far, we believe that this route has potential to upgrade bio-liquids enriched in carboxylic acids. Furthermore, ketonization can also condense typical components of bio-oils like esters, [86–88] and, unlike zeolite upgrading, this reaction can be efficiently carried out under moderate amounts of water. [89]

4.4.3 AQUEOUS-PHASE PROCESSING OF BIOMASS DERIVATIVES

As indicated in Fig. 4, aqueous solutions of sugars, derived from the carbohydrate fraction of lignocellulose (i.e., cellulose and hemicellulose), can be used to produce second generation ethanol fuel through fermentation routes. Alternatively, these sugars can be transformed into a variety of useful derivatives by means of chemical and biological processes. [90,91] As will be addressed in this section, sugars and some of their derivatives can be catalytically processed in the aqueous phase to produce liquid fuels chemically identical to those currently used in the transportation sector. The key advantage of this route, in comparison with BTL and pyrolysis-upgrading approaches, is derived from the mild reaction conditions used, allowing for better control of conversion selectivity. However, costly pre-treatment and hydrolysis steps are required to hydrolyze solid lignocel-lulose to soluble sugar feeds, and the lignin fraction, once isolated, is typi-cally combusted to provide heat and power.

The production of liquid hydrocarbon transportation fuels from bio-mass derivatives involves deep chemical transformations. In this respect, sugars (and chemicals derived from them) are molecules with high degrees of functionality (e.g., $-OH$, $-C{=}O$ and $-COOH$ groups) and a maximum number of carbon atoms limited to six (derived from glucose monomers). On the other hand, hydrocarbon fuels are larger (up to C_{20} for diesel ap-plications) and completely unfunctionalized compounds. Consequently, a number of reactions involving oxygen removal (e.g., dehydration, hy-drogenation, and hydrogenolysis), combined with C–C coupling (e.g., al-dol condensation, ketonization, and oligomerization), will be required to convert sugars into hydrocarbon transportation fuels, and aqueous-phase catalytic processing offers the opportunity to selectively carry out those transformations. Importantly, two aspects are crucial to ensure econom-ic feasibility of the aqueous-phase route: (i) reduction of the number of processing steps by means of catalytic coupling approaches [92] and (ii) deoxygenation of biomass feedstocks with minimal consumption of hy-drogen from an external source. [93]

The main aqueous-phase routes to upgrade sugars and derivatives into liquid hydrocarbon transportation fuels are schematically shown in Fig. 6. The biomass derivatives have been selected in view of their potential to produce liquid hydrocarbon fuels. First, we will describe the catalytic route designed to convert glycerol into liquid hydrocarbon fuels. This route involves the integration of two processes: aqueous-phase reforming (APR) of glycerol to syngas and F–T synthesis. This approach is particu-larly interesting because glycerol is produced in large amounts as a waste stream of the growing biodiesel industry. [94] Furthermore, glycerol can be co-produced, along with ethanol, by bacterial fermentation of sugars [95] (Fig. 4). Secondly, we will address furfural and hydroxymethylfurfural (HMF) as important compounds obtained by chemical dehydration of biomass-derived sugars. Furfural and HMF can be used as platform chemicals for green diesel and jet fuel production through dehydration, hydrogenation and aldol-condensation reactions. More recently, our group has developed a two-step (involving sugar reforming/reduction and C–C coupling processes) cascade catalytic approach to convert aqueous solutions of sugars and polyols into the full range of liquid hydrocarbon fuels, and this process will be described in Section 4.3.3. Finally, we will

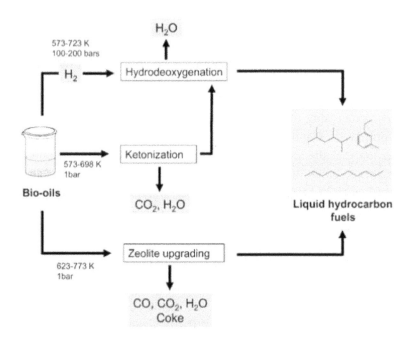

FIGURE 5: Catalytic routes for the upgrading of biomass-derived oils into liquid hydrocarbon transportation fuels.

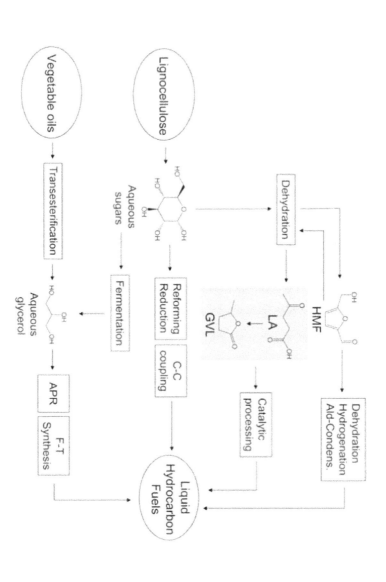

FIGURE 6: Main catalytic routes for the aqueous-phase conversion of sugars and derivatives into liquid hydrocarbon transportation fuels.

analyze the potential of two important biomass derivatives, levulinic acid (LA, obtained from sugars or HMF through dehydration processes) and g-valerolactone (GVL, obtained by hydrogenation of LA), to produce liquid hydrocarbon fuels.

4.4.3.1 GLYCEROL CONVERSION TO LIQUID HYDROCARBON FUELS.

Glycerol (1,2,3-propanetriol) is a water-soluble biomass-derived polyol with versatile chemical reactivity. [94,96,97] Approximately 100 kg of glycerol per ton of biodiesel in the form of concentrated aqueous solutions (80 wt%) are annually produced worldwide in the biodiesel industry as a byproduct of the transesterification of vegetable oils and animal fats with methanol. [98] The high number of applications of glycerol in different fields such as cosmetics, pharmaceuticals, foods, and drinks cannot absorb the surplus of this compound created by the biodiesel industry. Extra glycerol could be consumed as a fuel in internal combustion engines but, unlike ethanol, glycerol cannot be used blended with gasoline because of its low solubility in hydrocarbons and high viscosity. Furthermore, the purification of crude aqueous glycerol for chemical purposes is costly and energy-consuming. Consequently, new technologies, with potential to upgrade aqueous solutions of glycerol, would be valuable. In this respect, a promising route for glycerol conversion involves the production of syngas through aqueous-phase reforming (APR) processes. [99] By means of this route, concentrated aqueous solutions of glycerol can be converted to gaseous H_2, CO and CO_2 mixtures over supported metal catalysts at moderate temperatures (498–548 K). Platinum is the preferred metal for this conversion because it favors C–C cleavage reactions (leading to CO, H_2 and CO_2) versus C–O cleavage reactions (leading to light hydrocarbons). [73,100] To selectively produce syngas versus CO_2-enriched mixtures, reaction conditions and catalytic materials must be selected to control WGS reactions. Thus, by operating with inert materials like carbon as a support (instead of using inorganic oxide supports that can activate water), low total pressures (to avoid high partial pressures of water), and high glycerol concentrations (allowing water to become the limiting reagent in WGS), syngas streams with adequate H_2/CO ratios for Fischer–Tropsch synthesis can be produced from aqueous glycerol over Pt/C catalysts. [99]

To allow for coupling of endothermic glycerol reforming with exothermic Fischer–Tropsch synthesis in a single reactor, it is crucial that the former process is carried out efficiently at low temperatures. However, under these conditions, the metallic surface of Pt is likely to be saturated with adsorbed CO, [101] thus decreasing the overall catalytic rate. One effective strategy to overcome this limitation involves the utilization of alloys such as Pt–Re and Pt–Ru, on which the strength of adsorption of CO is lower compared to Pt. [102,103] This new route involving low-temperature gasification of aqueous glycerol to syngas integrated with Fischer–Tropsch synthesis represents an interesting alternative to complex BTL approaches (Section 4.1). Thus, unlike biomass gasification, glycerol reforming can be carried out at temperatures within the range employed for Fischer–Tropsch synthesis, thereby allowing effective integration of both processes in a single reactor [104] with improved thermal efficiency (since heat required for endothermic reforming is provided by exothermic F–T process). Furthermore, concentrated aqueous solutions of glycerol (as produced in biodiesel facilities) can be converted to undiluted and impurity-free syngas, thereby eliminating the need for large gasifiers with oxygen-production plants and expensive gas-cleaning units. Thus, unlike BTL, this aqueous-phase route allows for costcompetitive operations at small scale, which is advantageous for the processing of distributed biomass resources.

4.4.3.2 HMF AND FURFURAL PLATFORMS TO HYDROCARBON FUELS.

Furan compounds such as furfural (2-furaldehyde) and HMFare obtained by dehydration of sugars over mineral acid catalysts such as HCl or H_2SO_4 at moderate temperatures (e.g., 423 K). These furan compounds find applications as chemical intermediates in the production of industrial solvents, polymers and fuels additives. Furfural is obtained by dehydration of C_5 sugars like xylose in a well-developed industrial process. [105] On the other hand, the large-scale production of HMF from C_6 sugars is more complicated, and several aspects still remain a challenge, one of which is the utilization of glucose as a sugar feedstock. Thus, HMF is typically produced from glucose with low yields, and current technologies include an additional isomerization step to fructose, since dehydration of fructose to HMF takes place with better selectivity and higher rates. [106,107] Furthermore, the control over the unwanted side reactions involving the reactant, intermediates and the final HMF product is critical. It is particularly important to prevent HMF from overreacting in the aqueous phase, and utilization of biphasic reactors where HMF is continuously extracted into an organic solvent has shown promising results. [108]

Furfural and HMF can be used as building blocks for the production of linear hydrocarbons (in the molecular weight appropriate for diesel and jet fuel) by means of a cascade process involving dehydration, hydrogenation and aldol-condensation reactions, [109,110] as shown in Fig. 7 for HMF. The process starts with acid-catalyzed deconstruction of polysaccharides (e.g., starch, cellulose or hemicellulose) to yield C_5 and C_6 sugar monomers, which are subsequently dehydrated (under the same acid environment) to form carbonyl-containing furan compounds such as furfural and HMF. In a further step, the carbonyl group in the furan compounds serves as a reactive center for C–C coupling through aldol condensation reactions with carbonyl-containing molecules such as acetone, which can be also obtained from biomass-derived sources. [111,112] These condensations are base-catalyzed (e.g., NaOH, Mg–Al oxides) and are typically carried out in polar solvents like water. As a result of the aldol-condensation, a larger compound containing unsaturated C==C and C==O bonds (i.e., aldol-adduct) is formed and, owing to its hydrophobic character, this adduct precipitates out of the aqueous solution. Recently, improvements have been made in the aldol-condensation process by utilizing biphasic reactors where furan compounds (dissolved in organic THF) are contacted with aqueous NaOH, thus allowing continuous extraction of aldol-adducts into the organic phase. [110] As represented in Fig. 7, the molecular weight of final alkanes can be increased by allowing adducts to undergo a second aldolcondensation process with the initial furanic feedstock. The unsaturated C==C and C==O bonds in aldol adducts are subsequently hydrogenated over metal catalysts such as Pd to yield large water-soluble polyol compounds. The complexity of the process can be reduced by using a bifunctional (metal and basic sites) water-stable Pd/MgO–ZrO$_2$ catalyst. [113] Thus, both aldolcondensation and adduct hydrogenation can be carried out simultaneously in a single reactor. The last step of the process involves complete oxygen removal from the hydrogenated aldoladducts to produce liquid alkanes through aqueous-phase dehydration/hydrogenation (APD/H) reactions. [114] Oxygen is progressively removed from the water-soluble adducts over a bifunctional metal–acid catalyst by cycles of dehydration and hydrogenation reactions. APD/H can be achieved over Pt–SiO$_2$– Al$_2$O$_3$ in a four-phase reactor involving aqueous solution of adducts, a hydrogen gas inlet stream, a hexadecane sweep stream, and the solid catalyst. [109] The hexadecane stream is important in that it prevents intermediate organic species from overreacting to coke over acid sites. Recently, the utilization of a bifunctional Pt/NbPO$_4$ catalyst, with superior dehydration activity under water environments, [115] has allowed elimi-

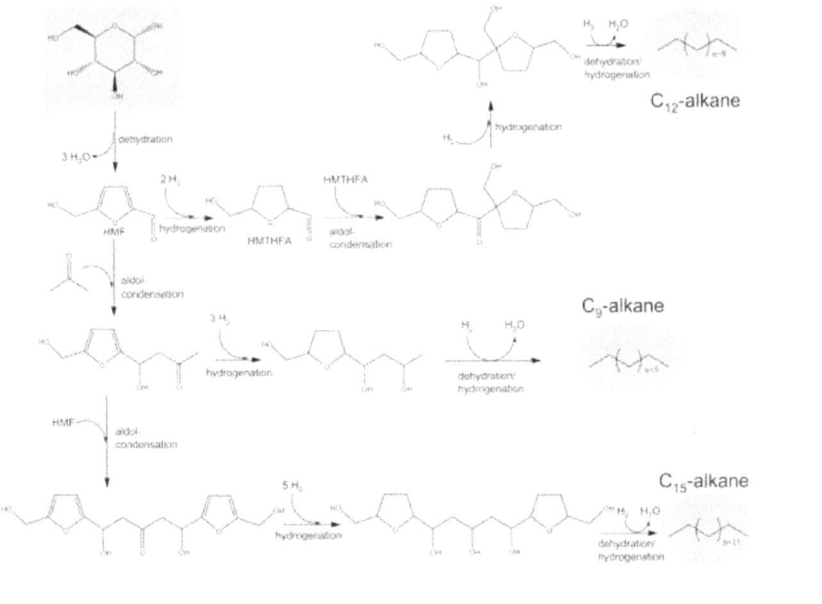

FIGURE 7: Reaction pathways for the conversion of biomass-derived glucose into liquid alkanes via HMF. Adapted from ref. 109.

nation of the hexadecane sweep stream step and, consequently, production of a pure organic stream of liquid hydrocarbon fuels with targeted molecular weights (C_9–C_{15} for HMF and C_8–C_{13} for furfural) that spontaneously separates from water and retains 60% of the carbon of the initial sugar feedstock. [110]

4.4.3.3 REFORMING/REDUCTION OF SUGARS.

The catalytic transformation of sugars to liquid hydrocarbon fuels is a complicated process that ideally should combine deep oxygen removal and adjustment of the molecular weight using a small number of reactors and with minimal utilization of fossil fuel-based external hydrogen. This goal can be achieved by (i) using multifunctional catalysts able to carry out different reactions in the same reactor [92] and (ii) utilizing a fraction of the sugar feedstock as a source of in situ hydrogen through aqueous-phase reforming reactions. [73]

Both approaches are combined by Kunkes et al. in a recent process that transforms aqueous solutions of sugars and sugaralcohols into liquid hydrocarbon fuels in a two-step cascade process [116] (Fig. 8). Firstly, aqueous sugars and polyols (typically glucose and sorbitol) are converted into a mixture of monofunctional compounds (e.g., acids, alcohols, ketones and heterocycles) in the C_4–C_6 range, which are stored in an organic phase that spontaneously separates from water. This step is carried out at temperatures near 500 K over a Pt–Re/C catalyst, which achieves deep deoxygenation (up to 80% of the oxygen in the initial feedstock is removed) by means of C–O hydrogenolysis reactions. Importantly, the hydrogen required to accomplish the C–O cleavage step is internally supplied by aqueous-phase reforming (involving C–C cleavage and WGS reactions) of a fraction of the feed (Fig. 8). The Pt–Re/C material allows production of hydrogen and removal of oxygen in a single reactor. Unlike bio-oils produced by pyrolysis (Section 4.2), the organic stream of monofunctional compounds produced by sugar processing over Pt–Re/C is completely free of water and has a well-defined composition that is controlled by the feedstock type (e.g., sugars or polyols) and the reaction conditions. [117]

The retention of functionality in the organic intermediates is key to control reactivity and to allow subsequent C–C coupling upgrading strategies. This approach has been demonstrated to be conceptually adequate to process sugars into fuels, [3] and important biomass derivatives such as lactic acid (3-hydroxypropanoic acid) [89,118] and levulinic acid [119] have been upgraded following this strategy. Each group of compounds (e.g., alcohols, ketones, acids) in the monofunctional stream can be upgraded to targeted hydrocarbons through different C–C coupling reactions (e.g., oligomerization, aldol-condensation and ketonization). For example, the organic stream enriched in alcohols by hydrogenation of ketones can be processed over an acidic H-ZSM5 zeolite at atmospheric pressure to yield 40% of C_{6+} aromatic gasoline components. Ketones can be upgraded to larger hydrocarbon compounds ($C_8–C_{12}$) with low extents of branching by means of aldol-condensation reactions over bifunctional Cu/ $Mg_{10}Al_7O_x$ catalysts. However, carboxylic acids present in the organic stream caused deactivation of the basic sites responsible for aldol-condensation, and approaches based on upstream removal of acids by ketonization (similar to those proposed herein for bio-oils upgrading, Section 4.2) and subsequent aldolcondensation have been successfully developed. [120–122] Ketonization acquires special relevance when the organic stream is rich in carboxylic acids, as is the case when the feed is glucose.

4.4.3.4 LEVULINIC ACID AND G-VALEROLACTONE PLATFORMS TO HYDROCARBON FUELS.

Levulinic acid (4-oxopentanoic acid) is an important biomass derivative that can be obtained by acid hydrolysis of lignocellulosic wastes, such as paper mill sludge, urban waste paper, and agricultural residues, through the Biofine process. [123] Levulinic acid has been recently selected by the US Department of Energy (DOE) as one of the top 15 carbohydrate-derived chemicals in view of its potential to serve as a building block for the development of biorefinery processes. [91] Thus, taking advantage of its dual functionality (i.e., a ketone and an acid group), a number of useful chemicals can be synthesized from levulinic acid including methyltetrahydrofuran (MTHF) (a gasoline additive) and d-aminolevulinic acid (DALA) (a biodegradable pesticide). [124]

Recently, our group has developed a series of catalytic approaches to convert aqueous solutions of levulinic acid into liquid hydrocarbon transportation fuels of different classes (Fig. 9). The catalytic pathways involve oxygen removal by dehydration/hydrogenation (in the form of water) and decarboxylation (in the form of CO_2) reactions, combined with C–C coupling processes such as ketonization, isomerization, and oligomerization that are required to increase the molecular weight and to adjust the structure of the final hydrocarbon product. As a first step, aqueous levulinic acid is hydrogenated to water-soluble GVL, which is the key intermediate for the production of hydrocarbon fuels. This hydrogenation step can be achieved with high yields by operating at low temperatures (e.g., 423 K) over non-acidic catalysts (e.g., Ru/C) to avoid formation of angelica lactone, a known coke precursor [125] which is produced by dehydration over acidic sites at higher temperatures (e.g., 573–623 K). [119] Interestingly, because equimolar amounts of formic acid (a hydrogen donor) are coproduced along with levulinic acid in the C_6-sugars dehydration process, this hydrogenation step could be potentially carried out without utilizing hydrogen from an external source, and several groups have already explored this possibility. [126,127] This route is promising in that GVL has applications as a gasoline additive, [128] and as a precursor to polymers [129] and fine chemicals. [130]

Aqueous solutions of GVL can be upgraded to liquid hydrocarbon fuels by following two main pathways: the C_9 route and the C_4 route (Fig. 9). In the former route, GVL is converted to 5-nonanone over a water-stable multifunctional Pd/Nb_2O_5 catalyst. In this process, GVL is first transformed into hydrophobic pentanoic acid by means of ring-opening (on acid sites) and hydrogenation reactions (on metal sites) at moderate temperatures and pressures. Pentanoic acid is subsequently ketonized to 5- nonanone, and reaction conditions can be adjusted to allow this transformation to take place on the same Pd/Nb_2O_5 reactor with a maximum of 70% carbon yield. [119] Nonanone yield can be increased to almost 90% by using a dual-catalyst approach with $Pd/Nb_2O_5 + Ce_{0.5}Zr_{0.5}O_2$ in a reactor with two different temperature zones, which allows for optimum control of reactivity. [131]

5-Nonanone, which is obtained in a high purity organic stream that spontaneously separates from water, is subsequently transformed into its corresponding alcohol that serves as a platform molecule for the production of hydrocarbon fuels for gasoline and diesel applications. For example, the C_9 alcohol can be processed (through hydrogenation/dehydration cycles) over a bifunctional metal–acid catalyst such as Pt/Nb_2O_5 [110] into linear n-nonane, with excellent cetane number and lubricity to be used as a diesel blender agent. Alternatively, the functionality of 5-nonanol can be utilized to upgrade the alcohol to gasoline and diesel components. In particular, 5-nonanol can be dehydrated and isomerized in a single step over an USY

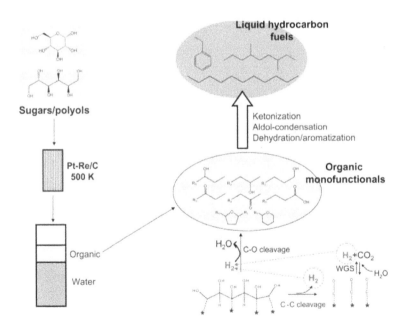

FIGURE 8: Scheme of the process for the catalytic conversion of sugars and polyols into liquid hydrocarbon fuels. Sugars primarily undergo reforming/reduction over Pt–Re/C to generate intermediate hydrophobic monofunctionals. The intermediates can be upgraded to liquid hydrocarbon fuels by means of C–C coupling reactions. Adapted from ref. 116.

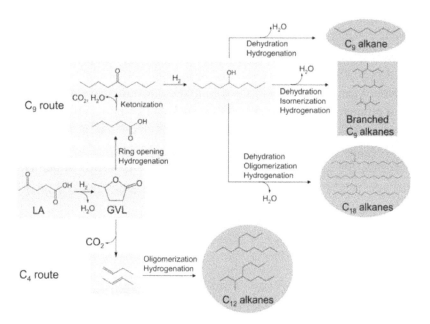

FIGURE 9: Catalytic routes for the conversion of levulinic acid (LA) and g-valerolactone (GVL) into liquid hydrocarbon transportation fuels. Blue colour indicates water-soluble compounds, yellow symbolizes hydrophobic compounds, and green refers to liquid hydrocarbon fuels.

zeolite catalyst to produce a mixture of branched C_9 alkenes with the appropriate molecular weight and structure for use in gasoline after hydrogenation to the corresponding alkanes. [131] Additionally, 5-nonanol can be converted into a C_9-alkene stream (by means of dehydration reactions) which can be subsequently oligomerized over an acid catalyst such as Amberlyst 70 to achieve good yields of C_{18} alkanes (after hydrogenation) for diesel applications. [132]

Recently, a promising route to upgrade aqueous solutions of GVL into jet fuels through the formation of C_4 alkenes has been developed by Bond et al. [133] (Fig. 9). The process is based on a dual reactor system. In the first catalytic reactor the GVL feed undergoes decarboxylation at elevated pressures (e.g., 36 bars) over a silica/alumina catalyst, producing a gas stream composed of butene isomers and CO_2. In a second reactor connected in series, the gaseous butene stream is passed over an acidic catalyst (H-ZSM5, Amberlyst 70) that achieves oligomerization of butene monomers, yielding

TABLE 1: Summary of the different technologies for the conversion of biomass into liquid hydrocarbon transportation fuels

	Technology						
	Thermal routes			Aqueous-phase routes			
	BTL	Pyrolysis	Glycerol reforming	HMF platform	Reforming of sugars	C9 route	GVL platform (C4 route)
Use of entire lignocellulose	Yes	Yes	No	No	No	No	No
Pretreatment required	Drying, size reduction	Drying, size reduction	Oil extraction	Pretreatment [acid hydrolysis dehydration]a, isomerization	Pretreatment, hydrolysis	Pretreatment [acid hydrolysis dehydration]a, hydrogenation	Pretreatment [acid hydrolysis dehydration]a, hydrogenation
External chemicals required	Oxygen	Inert gas	Methanol	Acetone, mineral acid, extracting solvent	Enzymes/mineral acid	Mineral acid	Mineral acid
Reaction conditions	1100–1500 K, 1–20 bar	650–800 K, 1 bar	498–548 K, 1–20 bar	298–393 K, 1–55 bar	483–523 K, 18–27 bar	598–698 K, 14–35 bar	443–648 K, 36 bar
Intermediate fraction	Syngas	Bio-oil	Syngas	Hydrogenated aldol-adduct	Monofunctionals organic stream	5-Nonanoneenriched organic phase	Butene + CO_2
Cleaning/conditioning/separation	Deep cleaning, WGS	Char removal	Not required	Extraction with organic solvent	Not required (spontaneous separation)	Not required (spontaneous separation)	Water removal
Upgrading process	F–T	HDO or zeolite upgrading	F–T	APD/H	C–C coupling, dehydration, hydrogenation	Hydrogenation, C–C coupling, dehydration	Oligomerization, hydrogenation

TABLE 1: *Cont.*

	Technology						
	Thermal routes		Aqueous-phase routes				
	BTL	Pyrolysis	Glycerol reforming	HMF platform	Reforming of sugars	GVL platform C9 route	GVL platform C4 route
Number of reactors	2–3b	2	2	4	4	5–7c	4
Precious metal catalysts involved	No	Yes	Yes	Yes	Yes	Yes	No
External H_2 requirements	No	High with HDO, none with zeolite	No	Moderate	No	Moderate	Minimum
LHF yield (g fuel per g dry biomass)	0.21d	0.45–0.15e	0.011f	0.27–0.10g	0.23–0.16h	0.25–0.19i	0.25–0.19j

a [ ] indicates that hydrolysis and dehydration can be carried out in the same reactor. b 3 reactors if WGS syngas conditioning is required. c Depending on the upgrading process required. d According to ref. 134. e Calculated as: [0.50–0.70 yield of bio-oil from lignocellulose in pyrolysis135] <?> [0.65 yield of LHF in HDO.136 or 0.30 yield of aromatics LHF in zeolite upgrading79]. f Calculated as: [0.20 content of oil in soybeans137] <?> [0.1 glycerol in biodiesel process] <?> [0.552 yield of alkanes104]. g Calculated as: [0.80–0.60 sugar content of lignocellulose] <?> [0.96 yield of hydrolysis enzymatic] <?> [0.53–0.31 yield isomerization glucose to fructose138,139] <?> [0.69–0.58 yield of LHF from fructose110]. h Calculated as: [0.80–0.60 sugar content of lignocellulose] <?> [0.96 yield of hydrolysis enzymatic] <?> [0.52 yield of organic C116] <?> [0.57 yield of C7+ ketones116]. i Yield to n-nonane (diesel blender). Calculated as: [0.80–0.60 sugar content of lignocellulose] <?> [0.96 yield of hydrolysis enzymatic] <?> [0.45 yield to levulinic acid of biofine process] <?> [0.96 yield of levulinic acid to GVL119] <?> [0.80 yield of 5-nonanone from GVL119] <?> [1.00] hydrogenation of 5-nonanone to n-nonane. j Calculated as: [0.80–0.60 sugar content of lignocellulose] <?> [0.96 yield of hydrolysis enzymatic] <?> [0.45 yield to levulinic acid of biofine process] <?> [0.96 yield of levulinic acid to GVL119] <?> [0.78 yield of C8+ alkenes133] <?> [1.00] hydrogenation to final alkane product.

a distribution of alkenes centred at C_{12}. While CO_2 does not affect the oligo-merization process other than by dilution, water inhibits the acidic oligo-merization catalyst (especially Amberlyst), and it has to be removed prior to the second reactor by using a gas–liquid separator operating at 36 bars and 373–398 K. The final optimized yield to C_{8+} alkenes reaches 75% when silica/alumina and Amberlyst 70 are used.

Since GVL can be potentially produced from levulinic acid with no external hydrogen requirements, this technology allows the production of liquid hydrocarbon alkanes from lignocellulose with minimal utilization of hydrogen (i.e., hydrogen is only used during the final alkene hydro-genation step). Furthermore, the process is potentially cost-competitive with petroleum-derived technologies, since only two reactors are required, operating in series and using non-precious metalcatalysts. Finally, a CO_2 gas stream is produced with high purity and at high pressures, thereby permitting effective utilization of sequestration or capture technologies to mitigate greenhouse gas emissions.

4.5 COMPARISON OF TECHNOLOGIES FOR BIOMASS CONVERSION INTO LIQUID HYDROCARBON FUELS

Table 1 summarizes the main characteristics of the different technologies analyzed in the present perspective paper for the conversion of biomass into liquid hydrocarbon fuels. Various routes have been compared in terms of important parameters (e.g., pretreatment required, external chemicals and hydrogen requirements, reaction conditions, cleaning/separation steps, number of reactors, use of precious metal catalysts) involved in the processing from the initial lignocellulosic biomass to the final liquid hy-drocarbon fuel. Additionally, technologies are compared in terms of the overall yield of LHF for each route, calculated according to recent data available in literature. The ability to use the entire organic matter in lig-nocellulose represents the main advantage of thermal routes (i.e., BTL and pyrolysis + upgrading) versus aqueous-phase approaches which, as indicated in Section 4.3, only can process the sugar fraction of lignocel-lulose (typically 60–80% depending on the source [13]). This constraint in aqueous-phase routes has important consequences for the overall process:

(i) it limits the final yield to LHF and (ii) it negatively affects the economics since additional reactors for pretreatment/hydrolysis steps are required to solubilize the sugar feedstocks (Table 1). On the other hand, aqueous-phase processes are carried out at milder conditions compared to thermal routes which allows for better control of the chemistry and, with it, higher selectivities to targeted hydrocarbon fuels. This control over the chemistry in aqueous phase technologies has important implications on the cleaning/ separation steps as well. Thus, unlike BTL and pyrolysis which produce intermediate fractions with high degrees of impurities that require deep cleaning and conditioning before the upgrading process, aqueous-phase routes achieve high-purity organic streams that spontaneously separate from water, with no further cleaning/ conditioning steps required. The higher number of reactors employed in the upgrading process for aqueous-phase routes versus thermal routes is another important difference between both approaches. In some cases (for example reforming/reduction of sugars, HMF and GVL platforms), the use of additional upgrading reactors is justified by the production of a final liquid hydrocarbon fuel with well-defined characteristics of molecular weight and structure that could be directly used for gasoline, jet fuel and diesel applications (see Fig. 7–9). This control of the final hydrocarbon fuel is more difficult to achieve by BTL and pyrolysis, which produce a mixture of hydrocarbons with broad molecular weight distributions and low control of the final chemical structure. Consequently, thermal routes might need additional refining reactors to produce fuel-grade compounds.

As indicated in previous sections, two parameters are important to assess the economic feasibility of aqueous-phase catalytic routes: the number of reactors and the use of external hydrogen. Thus, glycerol reforming, with only 2 reactors (reforming and F–T) and no hydrogen requirements (Table 1), would represent an interesting route. However, the final hydrocarbon yield (0.011 g of LHF per g of dry biomass), negatively affected by the low content of oils in the biomass source (20% in soybeans137), is low compared with other aqueous routes. With respect to these two parameters, reforming/reduction of sugars over Pt–Re is a promising route since no external hydrogen is required (hydrogen required for reduction of sugars and hydrogenation of the ketone formed by aldol-condensation or ketonization is internally supplied in sufficient amounts by aqueous-phase reforming of a fraction of the sugar, Fig. 8), and only 4 reactors (hydrolysis, reforming, C–C coupling and dehydration/hydrogenation) are needed to produce gasoline, diesel and jet fuel components from lignocellulosic biomass with an overall yield comparable to that of BTL (0.21 g of LHF per g of dry biomass). The main drawback of this approach is the high cost of the Pt–Re (10 wt%) reforming catalysts. The recently developed GVL platform to produce butene oligomers offers an attractive alternative with minimum external hydrogen utilization (required only during the final alkene hydrogenation step) and no precious metal catalysts, which should give this process a promising economic assessment. The HMF platform route can achieve good yields to LHF (with a maximum yield close to 0.3) at the expense of needing external chemicals such as acetone and organic solvents (typically produced from fossil fuels like petroleum) and moderate amounts of hydrogen to carry out APD/H. The GVL C_9 route offers versatility to produce gasoline and C_9–C_{27} diesel components with acceptable yields, but it would require multiple reactors to transform biomass into the final hydrocarbon fuel depending on the upgrading process used. Finally, we note that taking into consideration all the parameters as a whole, pyrolysis coupled with upgrading processes appears to be a promising route to convert lignocellulose into LHF with high yields and low complexity. While advances have been made recently in the pyrolysis step, challenges for this route are currently focused on the upgrading process, with particular emphasis on two crucial aspects: (i) designing strategies for the reduction of hydrogen consumption during HDO and (ii) development of hydrothermally stable catalysts (preferably without precious metals) with high resistance to sulfur and alkaline impurities typically present in bio-oils.

4.6 CONCLUSIONS

The production of liquid transportation fuels from renewable sources such as biomass is a promising route that can help to reduce our dependence on fossil fuels and to mitigate global warming effects. Ethanol, the most abundantly produced biofuel at the present time, suffers from low energy-density and compatibility issues with the existing transportation infrastructure, which is based on petroleum-derived liquid hydrocarbons. The current limitations of ethanol as a fuel can be overcome by developing cost-effective technologies that allow conversion of non-edible lignocellulosic biomass into liquid hydrocarbon fuels chemically identical to those currently used in the transportation sector. In this respect, several promising routes are currently being developed worldwide, including gasification of biomass to syngas coupled with F–T synthesis (BTL), pyrolysis integrated with bio-oils upgrading processes, and aqueous-phase catalytic processing of biomass-derived sugars and derivatives. BTL and pyrolysis-upgrading are thermochemical routes that allow utilization of all the organic matter in lignocellulose, and aqueousphase processing is a catalytic route designed to operate with water-soluble sugars (and platform chemicals derived from them). While aqueous-phase routes require that the lignocellulosic biomass be subjected to pretreatment/hydrolysis steps, these routes offer the opportunity to selectively carry out a variety of reactions to achieve the deep chemical transformations and C–C coupling reactions required when converting sugars into liquid hydrocarbons. To be economically viable, these aqueous-phase routes should be carried out with a small number of reactors and with minimum utilization of external fossil fuel-based hydrogen sources, as illustrated in the examples presented in the present paper.

REFERENCES

1. EIA Annual Energy Review, 2009, http://www.eia.doe.gov/aer/pdf/aer.pdf, accessed September 2010.
2. Eurostat, Statistical Aspects of the Energy Economy, 2008, issue number 55/2009, http://epp.eurostat.ec.europa.eu/cache/ITY_OFFPUB/KS-SF-09-055/EN/KS-SF-09-055-EN.PDF, accessed September 2010.
3. D. Simonetti and J. A. Dumesic, ChemSusChem, 2008, 1, 725.
4. Energy Information Administration, International Energy Outlook, 2009, http://www.eia.doe.gov/oiaf/ieo/pdf/0484%282009%29.pdf, accessed September 2010.
5. EIA International Energy Outlook, 2010, www.eia.doe.gov/oiaf/ieo, accessed September 2010.
6. BP Statistical Review of World Energy, 2010, http://bp.com/ statisticalreview, accessed September 2010.
7. UK Energy Research Centre, The Global Oil Depletion Report, 2009, http://www.ukerc.ac.uk/support/Global%20Oil%20Depletion, accessed September 2010.

8. Climate Change, Synthesis Report. Intergovernmental Panel on Climate Change, 2007, http://www.ipcc.ch/publications_and_data/ar4/syr/en/contents.html, accessed September 2010.

9. President Bush State on the Union Address, White House, 2007, http://usgovinfo. about.com/b/2007/01/23/bush-delivers-his-seventhstate-of-the-union-address.htm, accessed September 2010.

10. Directive 2003/30/EC of the European Union Parliament, Official Journal of the European Union, 2003, http://ec.europa.eu/energy/ res/legislation/doc/biofuels/en_final.pdf, accessed September 2010.

11. F. Kreith and D. Y. Goswami, in Handbook of Energy Efficiency and Renewable Energy, CRC Press: Taylor and Francis Group, Boca Raton, FL, 2007.

12. M. Graziani and P. Fornasiero, in Renewable Resources and Renewable Energy: A Global Challenge, CRC Press: Taylor and Francis Group, Boca Raton, FL, 2007.

13. D. L. Klass, in Biomass for the Renewable Energy, Fuels and Chemicals, Academic press, London, 1998.

14. J. Chheda, G. W. Huber and J. A. Dumesic, Angew. Chem., Int. Ed., 2007, 46, 7164.

15. A. J. Ragauskas, C. K. Williams, B. H. Davison, G. Britovsek, J. Cairney, C. A. Eckert, W. J. Frederick, J. P. Hallett, D. J. Leak, C. L. Liotta, J. R. Mielenz, R. Murphy, R. Templer and T. Tschaplinski, Science, 2006, 311, 484.

16. R. Rinaldi and F. Sch€uth, Energy Environ. Sci., 2009, 2, 610.

17. US National Science Foundation, in Breaking the Chemical and Engineering Barriers to Lignocellulosic Biofuels: Next Generation Hydrocarbon Biorefineries, 2008, http://www.ecs.umass.edu/biofuels/Images/Roadmap2-08.pdf, accessed September 2010.

18. Worldwatch Institute Center for American Progress, in American Renewable Path to Energy Security, 2006, http://www.worldwatch.org/files/pdf/AmericanEnergy.pdf, accessed September 2010.

19. International Energy Report, Energy Technology Essentials, Biofuel Production, 2007, http://www.iea.org/techno/essentials2.pdf, accessed September 2010.

20. Renewable Fuels Association, Ethanol Industry Outlook, 2010, http://www.ethanolrfa.org/page/-/objects/pdf/outlook/RFAoutlook2010_fin.pdf?nocdn¼1, accessed September 2010.

21. Worldwatch institute center for american progress, in Biofuels for Transport, Earthscan, London, 2007.

22. M. A. Martin, New Biotechnol., 2010, DOI: 10.1016/j.nbt.2010.06.010.

23. Biomass Research and Development Board, National biofuels action plan, 2008, http://www1.eere.energy.gov/biomass/pdfs/nbap.pdf, accessed September 2010.

24. EPA/DOE sponsored web site, http://www.fueleconomy.gov/feg/ flextech.shtml, accessed September 2010.

25. J. R. Regalbuto, Science, 2009, 325, 822.

26. Public Law 110–140, Energy Independence and Security Act of 2007, 2007, http:// energy.senate.gov/public/_files/getdoc1.pdf, accessed September 2010.

27. US Environmental Protection Agency report, Water Phase Separation in Oxygenated Gasoline, http://www.epa.gov/oms/regs/fuels/rfg/waterphs.pdf, accessed September 2010.

28. Astm D 4806, Standard Specification for Denatured Fuel Ethanol for Blending with Gasoline for Use as Automotive Spark Ignition Engine Fuel, ASTM International, West Conshohocken, PA, 2010, DOI:10.1520/D4806-10, www.astm.org.
29. H. Shapouri, J. A. Duffield and M. Wang, in The Energy Balance of Corn Ethanol: An Update (reportno. 814, Office of the Chief Economist), US Department of Agriculture, 2002, http://www.transportation.anl.gov/pdfs/AF/265.pdf, accessed September 2010.
30. Chemical Potential Energy, http://physics.info/energy-chemical, accessed September 2010.
31. C. N. Hamelinck, R. A. A. Suurs and A. P. C. Faaij, Biomass Bioenergy, 2005, 29, 114.
32. R. K. Perrin, in Ethanol and Food Prices—Preliminary Assessment, Faculty Publications: Agricultural Economics, University of Nebraska, 2008, http://digitalcommons.unl.edu/ageconfacpub/49, accessed September 2010.
33. D. L. Klass, in Biomass for the Renewable Energy and Fuels, Encyclopedia of Energy, ed. C. J. Cleveland, Elsevier, London, 2004.
34. R. D. Perlack, L. L. Wright, A. F. Turhollow, R. L. Graham, B. J. Stokes and D. C. Erbach, in Biomass as Feedstock for a Bioenergy and Bioproducts Industry: The Technical Feasibility of a Billion-ton Annual Supply, Oak Ridge National Laboratory, 2005, DOE/GO-102005–2135, http://feedstockreview.ornl.gov/pdf/billion_ton_vision.pdf, accessed September 2010.
35. J. P. Lange, Biofuels, Bioprod. Biorefin., 2007, 1, 39.
36. J. J. Bozell, Clean, 2008, 36, 641.
37. M. Stocker, Angew. Chem., Int. Ed., 2008, 47, 9200.
38. D. Martin-Alonso, J. Q. Bond and J. A. Dumesic, Green Chem., 2010, 12, 1493.
39. US Department of Energy, Feedstock Composition Glossary, 2005, http://www1.eere.energy.gov/biomass/feedstock_glossary.html#C, accessed September 2010.
40. P. Kumar, D. M. Barrett, M. J. Delwiche and P. Stroeve, Ind. Eng. Chem. Res., 2009, 48, 3713.
41. A. Carrol and C. Somerville, Annu. Rev. Plant Biol., 2009, 60, 165.
42. R. Kumar, G. Mago, V. Balan and C. E. Wyman, Bioresour. Technol., 2009, 100, 3948.
43. P. D€urre, Biotechnol. J., 2007, 2, 1525.
44. T. C. Ezeji, N. Qureshi and H. P. Blaschek, Curr. Opin. Biotechnol., 2007, 18, 220.
45. A. Schirmer, M. A. Rude, X. Li, E. Popova and S. B. del Cardayre, Science, 2010, 329, 559.
46. R. Luque, Energy Environ. Sci., 2010, 3, 254.
47. J. Han, H. Sun, H. Lou and X. Zheng, Green Chem., 2010, 12, 463.
48. I. Kubickova, M. Snare, K. Eranen, P. Maki-Arvela and D. Y. Murzin, Catal. Today, 2005, 106, 197.
49. UOP/ENI EcofiningTM Process, http://www.uop.com/objects/UOP_ENI_Ecofining_Process.pdf, accessed September 2010.
50. L. R. Lynd, C. Wyman, M. Laser, D. Johnson and R. Landucci, in Strategic Biorefinery Analysis: Analysis of Biorefineries, Technical Report, National Renewable Energy Laboratory, USA, 2002, http://www.nrel.gov/docs/fy06osti/35578.pdf, accessed September 2010.

51. B. Kamm, Angew. Chem., Int. Ed., 2007, 46, 5056.
52. G. W. Huber, S. Iborra and A. Corma, Chem. Rev., 2006, 106, 4044.
53. B. Kavalov and S. D. Peteves, in European Commission Joint Research Centre. Status and Perspectives of Biomass-to-Liquid Fuels in the European Union, 2005, http://www.mangus.ro/pdf/Stadiul%20actual%20si%20perspectivele%20bio-com-bustibililor%20in%20Europa.pdf, accessed September 2010.
54. T. A. Milne, R. J. Evans and N. Abatzoglou, in Biomass Gasifier Tars: Their Nature, Formation and Conversion, Report No. NREL/ TP-570–25357, National Renewable Energy Laboratory, 1998, http://www.nrel.gov/docs/fy99osti/25357.pdf, accessed September 2010.
55. H. Boerrigter and A. Van Der Drift, in Biosyngas: Description ofR&D Trajectory Necessary to Reach Large-Scale Implementation of Renewable Syngas from Biomass, Energy Research Centre of the Netherlands, 2004.
56. P. L. Spath and D. C. Dayton, in Preliminary Screening-Technical and Economic Assessment of Synthesis Gas to Fuels and Chemicals with Emphasis on the Potential for Biomass-Derived Syngas, United States Department of Energy, National Renewable Energy Laboratory, 2003, http://www.nrel.gov/docs/fy04osti/34929.pdf, accessed September 2010.
57. L. Caldwell, in Selectivity in Fischer–Tropsch Synthesis: Review and Recommendations for Further Work, 1980, http://www.fischer-tropsch.org/DOE/DOE_reports/81223596/pb81223596.pdf, accessed September 2010.
58. M. E. Dry, Catal. Today, 2002, 71, 227.
59. A. Steynberg, Fischer–Tropsch Technology, Studies on Surface Science and Catalysis, 2004, vol. 152, p. 1.
60. J. P. Lange, Catal. Today, 2001, 64, 3.
61. R. Zhang, K. Cummer, A. Suby and R. C. Brown, Fuel Process. Technol., 2005, 86, 861.
62. S. Koppatz, C. Pfeifer, R. Rauch, H. Hofbauer, T. Marquard- Moellensted and M. Specht, Fuel Process. Technol., 2009, 90, 914.
63. D. C. Elliott, D. Beckman, A. V. Bridgwater, J. P. Diebold, S. B. Gevert and Y. Solantausta, Energy Fuels, 1991, 5, 399.
64. D. Mohan, C. U. Pittman and P. H. Steele, Energy Fuels, 2006, 20, 848.
65. Y.-C. Lin and G. W. Huber, Energy Environ. Sci., 2009, 2, 68.
66. S. Czernik and A. V. Bridgwater, Energy Fuels, 2004, 18, 590.
67. D. C. Elliott, Energy Fuels, 2007, 21, 1792.
68. E. Furimsky, Appl. Catal., A, 2000, 199, 147.
69. J. Wildschut, F. H. Mahfud, R. H. Venderbosch and H. J. Heeres, Ind. Eng. Chem. Res., 2009, 48, 10324.
70. J. Wildschut, M. Iqbal, F. H. Mahfud, I. Melian Cabrera, R. H. Venderbosch and H. J. Heeres, Energy Environ. Sci., 2010, 3, 962.
71. S. R. A. Kersten, W. P. M. van Swaaij, L. Lefferts and K. Seshan, Options for Catalysis in the Thermochemical Conversion of Biomass into Fuels, in Catalysis for Renewables: From Feedstock to Energy Production, ed. G. Centi and R. A. van Santen, Wiley-VCH, Weinheim, 2007.
72. S. Czernik, R. French, C. Feik and E. Chornet, Ind. Eng. Chem. Res., 2002, 41, 4209.
73. R. D. Cortright, R. R. Davda and J. A. Dumesic, Nature, 2002, 418, 964.

74. R. R. Davda and J. A. Dumesic, Chem. Commun., 2004, 36.
75. C. Zhao, Y. Kou, A. A. Lemonidou, X. Li and J. A. Lercher, Chem. Commun., 2010, 46, 412.
76. K. Ramesh, N. Sharma and N. Bakhshi, Energy Fuels, 1993, 7, 306.
77. J. D. Adjaye, S. P. R. Katikameni and N. N. Bakhshi, Fuel Process. Technol., 1996, 48, 115.
78. A. G. Gayubo, A. T. Aguayo, A. Atutxa, R. Aguado and J. Bilbao, Ind. Eng. Chem. Res., 2004, 43, 2610.
79. T. R. Carlson, T. P. Vispute and G. W. Huber, ChemSusChem, 2008, 1, 397.
80. M. Renz, Eur. J. Org. Chem., 2005, 979.
81. K. M. Dooley, A. K. Bhat, C. P. Plaisance and A. D. Roy, Appl. Catal., A, 2007, 320, 122.
82. T. S. Hendren and K. M. Dooley, Catal. Today, 2003, 85, 333.
83. C. A. Gartner, J. C. Serrano-Ruiz, D. J. Braden and J. A. Dumesic, J. Catal., 2009, 266, 71.
84. C. A. Gartner, J. C. Serrano-Ruiz, D. J. Braden and J. A. Dumesic, Ind. Eng. Chem. Res., 2010, 49, 6027.
85. T. A. Milne, F. Aglevor, M. S. Davis, D. Deutch, and D. Johnson, in Development in Thermal Biomass Conversion, ed. A. V. Bridgewater and D. G. B. Boocock, Blackie Academic and Professional, London, UK, 1997.
86. R. Klimkiewicz, E. Fabisz, I. Morawski, H. Grabowska and L. Syper, J. Chem. Technol. Biotechnol., 2001, 76, 35.
87. M. Glinski, W. Szymanski and D. Lomot, Appl. Catal., A, 2005, 281, 107.
88. C. A. Gartner, J. C. Serrano-Ruiz, D. J. Braden and J. A. Dumesic, ChemSusChem, 2009, 2, 1121.
89. J. C. Serrano-Ruiz and J. A. Dumesic, Green Chem., 2009, 11, 1101.
90. T. Werpy and G. Petersen, in Top Value Added Chemicals from Biomass, Vol. 1— Results of Screening for Potential Candidates from Sugars and Synthesis Gas, US Dep. Energy, Off. Sci. Tech. Inf., 2004, http://www.nrel.gov/docs/fy04osti/35523.pdf, accessed September 2010.
91. J. J. Bozell and G. R. Petersen, Green Chem., 2010, 12, 539.
92. D. A. Simonetti and J. A. Dumesic, Catal. Rev., 2009, 51, 441.
93. J. C. Serrano-Ruiz, R. M. West and J. A. Dumesic, Annu. Rev. Chem. Biomol. Eng., 2010, 1, 79.
94. M. Pagliaro and M. Rossi, in Future of Glycerol, New Usages for a Versatile Raw Material, RSC publishing, Cambridge, 2nd edn, 2010.
95. C. S. Gong, J. X. Du, N. J. Gao and G. T. Tsao, Appl. Biochem. Biotechnol., 2000, 84, 543.
96. D. Gulen, M. Lucas and P. Claus, Catal. Today, 2005, 102–103, 166.
97. R. S. Karinen and A. O. I. Krause, Appl. Catal., A, 2006, 306, 128.
98. T. S. Wiinikainen, R. S. Karinen and A. O. I. Krause, Conversion of Glycerol into Traffic Fuels, in Catalysis for Renewables: From Feedstocks to Energy Production, ed. G. Centi and R. A. van Santen, Wiley-VCH, Weinheim, 2007.
99. R. R. Soares, D. A. Simonetti and J. A. Dumesic, Angew. Chem., Int. Ed., 2006, 45, 3982.
100. R. Alcala, M. Mavrikakis and J. A. Dumesic, J. Catal., 2003, 218, 178.

101. R. He, R. R. Davda and J. A. Dumesic, J. Phys. Chem. B, 2005, 109, 2810.
102. E. Christoffersen, P. Liu, A. Ruban, H. L. Skriver and J. K. Norskov, J. Catal., 2001, 199, 123.
103. J. Greely and M. Mavrikakis, Nat. Mater., 2004, 3, 810.
104. D. A. Simonetti, R. Hansen, E. L. Kunkes, R. R. Soares and J. A. Dumesic, Green Chem., 2007, 9, 1073.
105. K. J. Zeitsch, The Chemistry and Technology of Furfural and its Many by-Products, in Sugar Series, Elsevier, Amsterdam, 1st edn, 2000, vol. 13, pp. 34–69.
106. J. Chheda, Y. Roman-Leshkov and J. A. Dumesic, Green Chem., 2007, 9, 342.
107. C. Moreau, M. Belgacem and A. Gandini, Top. Catal., 2004, 27, 11.
108. Y. Roman-Leshkov, J. Chheda and J. A. Dumesic, Science, 2006, 312, 1933.
109. G. W. Huber, J. Chheda, C. Barret and J. A. Dumesic, Science, 2005, 308, 1447.
110. R. M. West, Z. Y. Liu, M. Peter and J. A. Dumesic, ChemSusChem, 2008, 1, 417.
111. D. T. Jones and D. R. Woods, Microbiol. Rev., 1986, 50, 484.
112. M. Sasaki, K. Goto, K. Tajima, T. Adschiri and K. Arai, Green Chem., 2002, 4, 285.
113. C. Barret, J. Chheda, G. W. Huber and J. A. Dumesic, Appl. Catal., B, 2006, 66, 111.
114. G. W. Huber, R. D. Cortright and J. A. Dumesic, Angew. Chem., Int. Ed., 2004, 43, 1549.
115. D. Braden, R. West and J. A. Dumesic, J. Catal., 2009, 262, 134.
116. E. L. Kunkes, D. A. Simonetti, R. M. West, J. C. Serrano-Ruiz, C. A. Gartner and J. A. Dumesic, Science, 2008, 322, 417.
117. R. M. West, E. L. Kunkes, D. A. Simonetti and J. A. Dumesic, Catal. Today, 2009, 147, 115.
118. J. C. Serrano-Ruiz and J. A. Dumesic, ChemSusChem, 2009, 2, 581.
119. J. C. Serrano-Ruiz, D. Wang and J. A. Dumesic, Green Chem., 2010, 12, 574.
120. E. L. Kunkes, E. Gurbuz and J. A. Dumesic, J. Catal., 2009, 266, 236.
121. E. Gurbuz, E. L. Kunkes and J. A. Dumesic, Green Chem., 2010, 12, 223.
122. E. Gurbuz, E. L. Kunkes and J. A. Dumesic, Appl. Catal., B, 2010, 94, 134.
123. S. W. Fritzpatrick, World patent, 9640609, Biofine Incorporated, 1997.
124. J. J. Bozell, L. Moens, D. C. Elliott, Y. Wang and G. G. Neuenscwander, Resour., Conserv. Recycl., 2000, 28, 227.
125. P. Ayoub and J. P. Lange, World patent, WO/2008/142127, 2008.
126. H. Heeres, R. Handana, D. Chunai, C. B. Rasrendra, B. Girisuta and H. J. Heeres, Green Chem., 2009, 11, 1247.
127. L. Deng, J. Li, D. M. Lai, Y. Fu and Q. X. Guo, Angew. Chem., Int. Ed., 2009, 48, 6529.
128. I. T. Horvath, H. Mehdi, V. Fabos, L. Boda and L. T. Mika, Green Chem., 2008, 10, 238.
129. J. P. Lange, J. Z. Vestering and R. J. Haan, Chem. Commun., 2007, 3488.
130. L. E. Manzer, Appl. Catal., A, 2004, 272, 249.
131. J. C. Serrano-Ruiz, D. J. Braden, R. M. West and J. A. Dumesic, Appl. Catal., B, 2010, 100, 184.
132. D. Martin-Alonso, J. Q. Bond, J. C. Serrano-Ruiz and J. A. Dumesic, Green Chem., 2010, 12, 992.
133. J. Q. Bond, D. Martin-Alonso, D. Wang, R. M. West and J. A. Dumesic, Science, 2010, 327, 1110.

134. D. Bianchi, in Biomass Catalytic Conversion to Diesel Fuel: Industrial Experience, Next Generation Biofuels, Bologna, 18 September 2009, www.ics.trieste.it/media/139853/df6504.pdf, accessed October 2010.
135. Biomass Technology Group (BTG), Fast pyrolysis, http:// www.btgworld.com/index.php?id¼22&rid¼8&r¼rd, accessed October 2010.
136. J. Wildschut, M. Iqbal, F. H. Mahfud, I. Melian-Cabrera, R.H.Venderbosch andH. J.Heeres,EnergyEnviron.Sci., 2010, 3, 962.
137. USDA Nutrient database, http://www.nal.usda.gov/fnic/foodcomp/cgi-bin/list_nut_edit.pl, accessed October 2010.
138. M. Moliner, Y. Roman-Leshkov and B. Davis, Proc. Natl. Acad. Sci. U. S. A., 2010, 107, 6164.
139. W. Aehle, in Enzymes in Industry: Production and Applications, Wiley-VCH, Weinheim, 2nd edn, 2004, p. 198.

CHAPTER 5

Catalytic Upgrading of Bio-Oil by Reacting with Olefins and Alcohols over Solid Acids: Reaction Paths via Model Compound Studies

ZHIJUN ZHANG, CHARLES U. PITTMAN, JR., SHUJUAN SUI, JIANPING SUN, AND QINGWEN WANG

5.1 INTRODUCTION

Production of renewable fuels and chemicals from lignocellulosic biomass has attracted increasing attention because of decreasing oil reserves, enhanced fuel demand worldwide, increased climate concerns, and the inherent conflict between food prices and converting edible carbohydrates to ethanol or plant oils to bio-diesel [1–4]. Bio-oils, the liquid products obtained from biomass fast pyrolysis or liquefaction, are regarded as promising renewable energy sources by the virtue of their environmentally friendly potential [5,6]. Nonetheless, several drawbacks of bio-oil severely limit its potential to replace or supplement high-grade transportation fuels. These include low heating values, high corrosiveness, high water

Catalytic Upgrading of Bio-Oil by Reacting with Olefins and Alcohols over Solid Acids: Reaction Paths via Model Compound Studies. © *Zhang Z, Pittman, Jr. CU, Sui S, Sun J, and Wang Q. Energies 6 (2013), doi:10.3390/en6031568. Licensed under a Creative Commons Attribution 3.0 Unported License, http://creativecommons.org/licenses/by/3.0/.*

content, thermal instability and immiscibility with hydrocarbon fuels etc. [7]. Thus, bio-oil has to be upgraded before using it as a fuel.

Numerous upgrading approaches to improve the bio-oil properties have been proposed, including hydrodeoxygenation, zeolite cracking, catalytic pyrolyses, steam reforming, and integrated catalytic processing such as combination of hydroprocessing and catalytic pyrolysis with zeolite catalysis [8–16]. However, these methods require temperatures from 300 to 800 °C where coke and tar easily form. This results in undesirable catalyst deactivation and reactor clogging. Hydrodeoxygenation can remove most of the oxygen present in bio-oil; but it requires high pressures and substantial amounts of hydrogen, which would negatively affect the economics of this process.

Alternatively, bio-oil can be partially refined to less hydrophilic, more combustible and more stable oxygen-containing organic fuels, where oxygen is not fully removed. Hydrogen is not employed or consumed and carbon is not lost as CO_2 in this approach. Ideally, this process would retain all of the bio-oil's original caloric value in the product. An example is esterification of bio-oil's carboxylic acids with alcohols, which also converts some ketones and aldehyde content to acetals. This can improve the chemical and physical properties of bio-oil. However, water is formed in the process. Excess alcohol use and water removal during reaction is required to drive these equilibrium reactions and their separation from the upgraded products should be considered [10,12]. We recently reported a promising approach, where bio-oil was converted into oxygen-containing fuels by reacting with added olefins over solid acid catalysts at low temperatures [17,18]. In this approach, acid-catalyzed esterification of bio-oil carboxylic acids by alcohols formed during olefin hydration, phenol alkylation, etherification, and hydration reactions of olefins occur simultaneously to convert carboxylic acids, phenolic compounds, alcohols and water into esters, alkylated phenols, ethers and alcohols, respectively. These products are less hydrophilic and have a higher fuel value. Water is removed instead of being generated. The hydroxyl groups present in bio-oil were removed and the fuel value of the product was enhanced. By also adding a co-reagent alcohol, serious phase separation of the hydrophilic bio-oil and hydrophobic olefin was reduced. In addition, esterification and acetal formation occur and their equilibria are further driven by the removal of the product water from these reactions by its addition across the added olefins [19]. The alcohols selected, including ethanol and butanols, can be obtained by biomass fermentation [20,21], and they are fuels themselves. C-4 alcohols are now major industrial targets for carbohydrate or cellulose fermentation to fuels routes. If they become major commodity fuels, a portion of that production could be directed to and leveraged towards bio-oil to fuels manufacturing processes. While this future cannot be foreseen now, options should now be developed for the future.

Converting alcohols from gasoline additives to bio-oil refining re-agents, which end up in the fuel, does not change their ultimate caloric contribution for fuel use. Olefin mixtures can be used. So, although ole-fins are consumed that may have other uses, olefins or olefin mixtures, whatever is cheaper, can be applied. For example, cheaper olefin mixtures can be obtained by pyrolysis of waste polyolefin base plastics. The total caloric content of the combined olefin and alcohol reagents remain within the refined upgraded products together with all of the original caloric con-tent of the raw bio-oil. However, olefin/alcohol acid-catalyzed upgrading will not, and is not intended, to produce "drop in" fuels for use in gasoline or most diesel motors. Furthermore, the product is not primarily intended for subsequent feeding into refinery processes to make gasoline due to its considerable oxygen content. But the oxygenated products can be blended with petroleum fuels or biodiesel liquids and might have promise for ap-plication in low temperature/high compression diesel engines requiring low cetane number fuels someday.

In our earlier work, undesirable catalyst deactivation [18] and decom-position [17,19] occurred during these upgrading processes due to bio-oil's complex composition, especially with substantial amounts of water present. A goal of this work is to develop and apply a more highly active catalyst with good hydrothermal stability for this process. A second goal is to more fully elucidate the complex competing reaction pathways in-volved in the acid-catalyzed refining of bio-oil with olefins plus alcohols. Bio-oil upgrading is exceptionally complex. This is because bio-oils are composed of a wide variety of oxygenated compounds (more than 300) and its chemical composition is feedstock and pyrolysis process dependent. However, all reported bio-oils are poorly defined mixtures of carboxylic acids, alcohols, aldehydes, esters, ketones, sugars or anhydrosugars, phenolic compounds, furans, water, a few cyclic hydrocarbons and multifunctional compounds such as hydroxyacetaldehyde, hydroxyacetic acid, hydroxyacetone, etc. [7]. Model compounds and their mixtures are often first employed to study the reaction steps involved in bio-oil upgrading processes [10,12,13]. For this work, we have selected phenol, water, acetic acid, acetaldehyde hydroxyacetone, D-glucose and 2-hydroxymethylfuran as typical bio-oil components and mixed them as a synthetic bio-oil. This composition contains a better representation of the types of compounds involved in the many reactions competing during refining and has allowed a more complete understanding of the reaction paths. Solid acid catalyzed reactions of 1-octene/1-butanol with this synthetic bio-oil were investigated in the liquid phase, respectively, over Dowex50WX2 (DX2), Amberlyst15 (A15), Amberlyst36 (A36), $Cs_{2.5}H_{0.5}PW_{12}O_{40}$ (an insoluble acidic heteropoly acid salt) and silica sulfuric acid (SSA) catalysts. All of these catalysts are reported water-tolerant strong acids. A short preliminary communication in Bioresource Technology [22] on this effort has appeared. The full study is reported here.

SSA is a superior proton source compared with many acidic solid supports, such as styrene/divinylbenzene sulfonic acid resins and Nafion-H [23]. SSA exhibited good activity and stability in preliminary catalytic upgrading of model bio-oils by simultaneous reactions with 1-butanol and 1-octene [22]. Reaction pathways were proposed, but, more systematic research was needed to examine upgrading feasibility and elucidate the complicated reaction mechanism. Thus, various olefins and alcohols are investigated in reactions with phenol/water (1:1) mixtures in this paper. Also, reactions of 1-octene with phenol, phenol/water, phenol/water/acetic acid, phenol/water/1-butanol, phenol/water/2-hydroxymethylfuran, phenol/water/D-glucose, phenol/water/hydroxyacetone and phenol/water/acetic acid/1-butanol are reported here. Herein, we present a more comprehensive reaction pathway and demonstrate coking/catalyst poisoning caused by hydroxyacetone, 2-hydroxy-methylfuran and D-glucose.

5.2 RESULTS AND DISCUSSION

5.2.1 CATALYST CHARACTERIZATION

The silica sulfuric acid (SSA), prepared by reacting silica gel with chlorosufonic acid in dichloromethane was obtained as a white solid in 98% yield. Table 1 summarizes the physical properties (surface area, pore size, pore volume and acidity amount) as well as chemical compositions of SSA and other four catalysts. The specific surface area was calculated using the BET equation. The total pore volume was determined at 77K for 300 min and also the average pore diameter were was calculated using the Barrett-Joyner-Halenda (BJH) method. The amount of H^+ was calculated by titration of catalyst samples in water with standard sodium hydroxide (0.495M). These results show that SSA have a good specific surface area than the three ion exchange resins and high pore volume than $Cs_{2.5}/K10$. This might be a reason for high catalytic performance of the SSA catalyst in the experimental conditions. Negligible decreases in pore volume, surface area and pore diameter of once used SSA catalyst (Table 1) displayed its good reusability.

Table 2 summarizes its typical IR absorptions and their assignments. The strong broad absorption bands from 1000 to 1100 cm^{-1} correspond to Si-O-Si bridge stretching vibrations (1097 and 1065 cm^{-1}) in silica [23]. The peak at ca. 971 cm^{-1} is associated with Si-OH stretching vibrations in silica. Bands appearing at ca. 852 and 886 cm^{-1} were assigned to the symmetrical and asymmetrical S-O stretching, respectively [24]. The peak at ca. 1178 cm^{-1} is the asymmetric S=O stretching vibration, while S=O symmetrical stretching vibrations lies at 1010–1080 cm^{-1}, overlapped by Si-O stretching bands [24]. The strong broad absorption at about 3200–3500 cm^{-1} is due to hydrogen bonded -OH in SSA. Characteristic IR absorptions of $Cs_{2.5}/K10$ are also summarized in Table 2. The IR bands at ca. 1075 cm^{-1}, 1032 cm^{-1} and 982 cm^{-1} were due to P-O in the central tetrahedron, K10 clay and terminal W=O, respectively. The peaks at ca.886 and 790 cm^{-1} (asymmetric W-O-W vibrations) are associated with the Keggin polyanion [18].

TABLE 1; Characteristics of catalysts[a].

Catalysts	Cs$_{2.5}$/K10	Amberlyst36	Amberlyst15	Dowex50X2	SSA
Description	Cs$_{2.5}$H$_{0.5}$PW$_{12}$O$_{40}$ supported on clay	Macroreticular resin	Macroreticular resin	Microreticular resin	Silica sulfuric acid
Chemical composition	Cs$_{2.5}$H$_{0.5}$PW$_{12}$O$_{40}$/aluminosilicate(lamellar), 30%	Functionalized copolymers (styrene + DVB)	Copolymer (styrene + 20% DVB)	Copolymer (styrene + 2% DVB)	SSA
Acidity type	Brönsted + Lewis	Brönsted	Brönsted	Brönsted	Brönsted
Acidity amount (meq·g^{-1})	0.17	5.4	4.7	4.3	2.9 (2.7)
BET surface area (m^2·g^{-1})	181	35	51	Gel (swells)	308 (302)
Average pore diameter (nm)	6.0	24	40-80	NA	2.1 (2.0)
pore volume (cm^3·g^{-1})	0.29	0.20	0.40	NA	0.509 (0.498)

[a] All the characteristics of the three resin catalysts were provided by the manufacturer; NA: not available; The corresponding characteristics of once used SSA (washed 3 times with acetone, dried in an oven at 105 °C for 30 min prior to test) are shown in parentheses as bold italic type.

TABLE 2. Characteristic IR absorptions for SSA and $Cs_{2.5}$/K10 catalysts [22].

SSA		$Cs_{2.5}$/K10	
Absorption (cm^{-1})	Assignment	Absorption (cm^{-1})	Assignment
3200~3500	hydrogen bonded -OH	1075	P-O stretching
1178	S=O symmetrical stretching	1032	Si-O in K10 clay
1000~1100	Si-O-Si bridge stretching	982	terminal W=O
971	Si-OH stretching 8	86,790	W-O-W vibrations
852, 886	S-O stretching		

5.2.2 CATALYTIC ACTIVITY

Phenol, water, acetic acid, acetaldehyde, hydroxyacetone, D-glucose and 2-hydroxymethylfuran were mixed together and used as a model bio-oil to react with 1-octene/1-butanol at 120 °C for 3 h over each of the five catalysts: $Cs_{2.5}$/K10, A15, A36, DX2 and SSA. Table 3 shows the 1-octene, 1-butanol and phenol conversions as well as the 1-octene isomerization and O-alkylation selectivities of these reactions [22]. 1-Octene conversions differed significantly over these catalysts and followed the order: SSA ($\approx$60%) > DX2 (40–50%) > A15 (27%) > A36 (14%) > $Cs_{2.5}$/K10 (10%). Similar differences occurred for both phenol conversion and 1-octene isomerization. The phenol conversion was higher with SSA (64.1%) verses DX2 (37.3%), A15 (27.6%), A36 (6.1%) and $Cs_{2.5}$/K10 (1.2%). The 1-octene isomerization activities of these five catalysts are 87.9% (SSA), 55.5% (DX2), 54.1% (A15), 13.5% (A36) and 1.9% ($Cs_{2.5}$/K10). These follow the same order and show the higher activity of the SSA catalyst. Higher phenol conversion was accompanied by higher 1-octene isomerization activity and higher 1-octene conversions. SSA is the most active catalyst. This is because it is a stronger acid than the three resin sulfonic acids. Compared with the resin sulfonic acids (P-C_6H_4-SO_3H), where the S atom has 3 O atoms attached, the S atom in SSA (SiO_2-OSO_3H) has 4 O atoms attached. This causes the weaker basicity of -O-SO_3^- verses that of Ph-SO_3^-. Thus, SSA is the strongest acid. The stronger the acid, the more 1-octene protonation is favored. Hence, more octyl cations are generated. With the increase in octyl cation concentration, both phenol alkylation (phenolic oxygen attack on the carbocation) and 1-octene isomerization reaction (loss of proton from the carbocation) would speed up accompanied with faster consumption of 1-octene. This is consistent with higher phenol conversion and both 1-octene isomerization activity and conversion to other products with SSA.

TABLE 3: 1-Octene, 1-butanol and phenol conversions, 1-octene isomerizations and O-alkylation selectivities in reactions of a model bio-oil with 1-octene/1-butanol over $Cs_{2.5}$/K10, A36, A15, DX2 and SSA at 120 °C for 3 h[a].

Catalyst	1-Octene conversion (%)[b]	Phenol conversion (%)[c]	1-Octene isomerization (%)[d]	1-Butanol conversion (%)[e]	O-alkylates selectivity (%)[f]
$Cs_{2.5}$/K10	10.1	1.2	1.9	68.2	42.0
A36	14.0	6.1	13.5	94.3	67.7
A15	27.1	27.6	54.1	97.4	73.7
DX2	43.1	37.3	55.5	90.5	73.9
SSA	60.1	64.1	87.9	97.4	64.1

[a]*Material ratio: 1-octene: 1-butanol: phenol: water: acetic acid: acetaldehyde: hydroxyacetone: D-glucose: 2-hydroxymethylfuran (g) = 1.35: 0.15: 0.94: 0.15: 0.15: 0.12: 0.12: 0.15: 0.15, catalyst: 0.15g;* [b]*1-Octene conversion = 100% × (1- GC area% of unreacted octenes versus the sum of the GC area% of alkylated phenols, octanols, dioctyl ethers, oligomers and octyl acetates);* [c]*Phenol conversions = 100% × GC area% of phenol alkylates versus the sum of the GC area% of unreacted phenol and phenol alkylates;* [d]*Percent of 1-octene isomerization = 100% × (1- GC area% of 1-octene versus the sum of the GC area% of 1-octene and 1-octene isomers);* [e]*1-Butanol conversion = 100% × (1- GC area% of unreacted 1-butanol versus the sum of the GC area% of butyl acetate, dibutyl ether, 1,1-dibutoxyethane and butyl levulinate);* [f]*O-Alkylates selectivity = 100% × GC area% of O-alkylates versus the sum of the GC area% of all phenol alkylates.*

Stronger acids also promote both esterification and acetal formation rates. This can be observed from the higher 1-butanol conversion (97.4%) with SSA catalyst. Except for the modest 1-butanol conversion (68%) formed over $Cs_{2.5}$/K10, 1-butanol conversions with the three resin sulfonic acids catalysts all exceeded 90%. Carboxylic acid esterifications and aldehyde/ketone acetal formation with 1-butanol are reversible or equilibrium reactions. The desirable forward reaction products like esters and acetals were produced accompanied by formation of water. That water and the original water present in bio-oil would inhibit the forward reactions, limiting further formation of more esters and acetals. Water removal by acid catalyzed hydration of 1-octene helped to shift these equilibria forming esters and acetals toward completion.

Phenol alkylates (C- and O-) are desired because of their high octane number and high heating values [15]. The O-alkylated products are especially desirable because the acidic phenolic hydroxyl group is converted to an ether lowering product acidity and decreasing hydrophilicity. Moreover, O-alkylated phenol ethers are readily combusted. Except for $Cs_{2.5}$/K10, all the catalysts gave high O-alkylation selectivity (>60%). Compared with the three resin sulfonic acids, SSA gives more C-alkylates because that stronger acid promotes conversion of O-alkylates into the thermodynamic phenol C-alkylates by enhancing O-alkylate protonation.

SSA exhibited a higher water-tolerance than other catalysts based on the model systems shown in Table 3. Desulfonation of the three resin sulfonic acids catalyst occurred progressively at 120 °C over time, leading to partial deactivation of these catalysts. $Cs_{2.5}$/K10 lost almost all catalytic activity.

5.2.2 REACTIVITIES OF MODEL BIO-OIL COMPONENTS

In order to prove the feasibility of this upgrading process, more clearly outline the complicated reaction mechanism and probe the causes for catalyst deactivation, additional model reactions were investigated. Figure 1 shows the phenol conversions of phenol/1-octene reactions at 120 °C for 3 h over all five catalysts both with and without water present. All the catalysts exhibit high activities in neat phenol/1-octene reactions based on the high phenol conversions (>80%). Water significantly lowered the phenol conversions of phenol/1-octene reactions over DX2 (42.1%), A36 (38.3%), A15 (15.8%) and $Cs_{2.5}$/K10 (19.9%). However, a good phenol conversion (74.5%) was still obtained over SSA. This further illustrated the high activity of SSA under hydrothermal conditions.

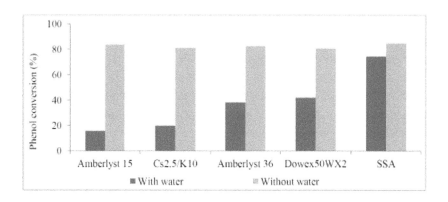

FIGURE 1: Comparison of phenol conversions by 1-octene with and without water present at 120 °C over Amberlyst15, $Cs_{2.5}/K10$, Amberlyst36, Dowex50WX2 and SSA (catalyst, 0.15 g; time: 3 h; 1-octene: phenol: water (mmol) = 10:10:10).

Phenolic compounds are present in bio-oils, primarily derived from lignin species. These acidic phenolic fractions are prone to oligomerization reactions with other bio-oil components [25]. Friedel-Crafts-type alkylations of phenol with 1-octene over solid catalysts leads to a mixture of O- and C-alkylated phenols (Table 4). Isomeric octyl phenyl ethers (O-alkylates) and octyl phenols (ortho/para-C-alkylates) were formed (Scheme 1), indicating that the 2-octyl carbocation undergoes 1, 2-hydride shifts to generate the 3- and 4-octyl cations in competition with O- and C-alkylation. O-Alkylation is faster than C-alkylation but O-alkylation is reversible and the initially generated O-alkylated products can be increasingly converted to C-alkylated (thermodynamic) products as a function of reaction conditions. All catalyst used gave high phenol conversions (Figure 1).

Water is the most abundant compound in raw bio-oil. It is difficult to remove due to its miscibility with hydrophilic thermolysis products present from cellulose and hemicellulose [5,6]. Phenol conversions from the reactions of 1-octene with water/phenol over solid acid catalysts were summarized in Figure 1. The lower phenol consumption with all catalysts when water was present is most likely due to water solvation of the sulfonic acid sites which lowers the Bronsted acidity or to mass transport effects due to phase separation. The sulfonic acid resins showed higher phenol conversions than $Cs_{2.5}/K10$. This could be due to swelling of resins. This swelling allows a distribution of all reactants to access a larger fraction of the internal acid sites of this macroreticular resin. However, partial Amberlyst15 decomposition occurred. Product distributions of these reactions are shown in Table 5. Obviously, competition between water and phenol for 1-octene occurred because 1-octanol and its isomers were formed by water uptake. Intermolecular reaction of these octanols further formed ethers (Scheme 2). Meanwhile, the significant increase in the concentration of octanols with increasing water concentration affirmed the water consumption by olefin hydration [18]. Olefin acid-catalyzed hydration removes water. This is the key reason for the success of this upgrading process.

TABLE 4: Product distributions of 1-octene reaction with neat phenol over different catalysts at 120 °C[a].

	Selectivity (%)		
Catalyst	O-alkylates	C-alkylates	Di-C-alkylates
Amberlyst 15	5.5	86.0	8.5
$Cs_{2.5}/K10$	3.2	65.9	30.9
Amberlyst 36	8.0	86.3	5.7
Dowex50WX2	33.5	63.2	3.3
SSA	37.2	34.4	28.4

[a]Reaction conditions: (solid catalyst, 0.15 g; time: 3 h; 1-octene: phenol (mmol) = 10:10.

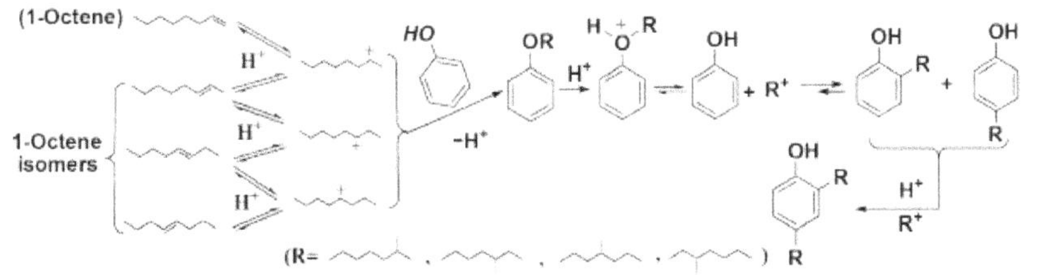

SCHEME 1: Acid-catalyzed reactions of phenol as a model phenolic compound with 1-octene.

TABLE 5: Product distributions from 1-octene reactions in phenol/water at 120 °C a.

Catalyst	Selectivity (%)[b]			Yield (%)	
	O-alkylates	C-alkylates	Di-C-alkylates	Octanols	Dioctyl ethers
Amberlyst 15	82.5	16.0	1.5	5.4	0.2
Cs$_{2.5}$/K10	75.9	20.8	3.3	6.7	0.4
Amberlyst 36	73.7	25.0	1.4	0.7	0.2
Dowex50WX2	84.5	12.5	3.0	4.7	1.7
SSA	75.1	19.5	5.4	3.4	1.7

[a]Reaction conditions: (solid catalyst, 0.15 g; time: 3 h; 1-octene: phenol: water (mmol) = 10:10:10. [b]GC area% of involved compounds versus the sum of the GC area% of all products that remained after the reaction.

Carboxylic acids, like acetic and propanoic acids, make bio-oil corrosive, especially at an elevated temperature [7]. 1-Octene was reacted with phenol/water/acetic acid solutions and their phenol conversions and octyl acetates yields were shown in Table 6. Along with the 1-octene hydration and phenol alkylation, simultaneous conversion of acetic acid to octyl acetates occurred by addition across 1-octene (Scheme 3), generating three groups of improved fuel components in one operation without water generation. Trace amount of phenyl acetate were formed. The phenyl acetate yield decreased with increasing temperatures from 60 to 100 °C [17]. Both SSA and DX2 show higher catalytic activity than other three catalysts based on related phenol conversions and octyl acetates yields.

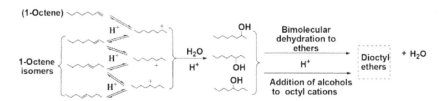

SCHEME 2: Acid-catalyzed reactions of water with 1-octene.

Table 6. Phenol conversions and octyl acetates yield of 1-octene reactions with phenol/water/acetic acid over different catalysts at 120 °C[a].

Catalyst	Phenol Conversion (%)	[b]Yield (%) Octyl acetates
Amberlyst 15	5.6	0.6
$Cs_{2.5}$/K10	2.3	0.4
Amberlyst 36	6.5	1.0
Dowex50WX2	63.1	4.3
SSA	67.2	9.4

[a]Reaction conditions: (solid catalyst, 0.15 g; time: 3 h; 1-octene: phenol: water: acetic acid (mmol) = 10:10:10:3.0; [b]GC area% of involved compounds versus the sum of the GC area% of all products remained after the reaction.

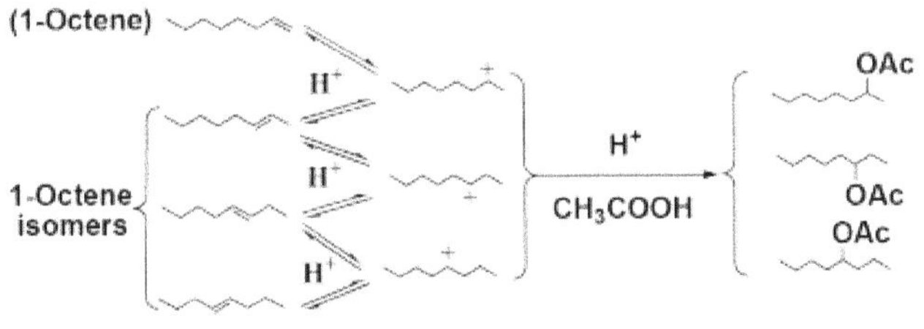

SCHEME 3: Acid-catalyzed reactions of acetic acid as a model carboxylic acid with 1-octene.

Bio-oil contains a large number of primary and secondary aliphatic hydroxyl groups from cellulose and hemicellulose pyrolysis. Table 7 provides the phenol conversions and butyl octyl ethers yields of 1-octene reactions with phenol/water/1-butanol mixtures over Dowex50WX2, SSA, Amberlyst15 and $Cs_{2.5}$/K10, respectively. Except for the reactions mentioned above, butyl octyl ethers were generated by either 1-butanol etherification with octanols or 1-butanol addition across octenes (Scheme 4). SSA gave the highest phenol conversion (69.2%), illustrating it had the highest catalytic activity. $Cs_{2.5}$/K10, which gave a phenol conversion of only 6.9%, lost almost all its activity during the reaction.

Multifunctional compounds such as hydroxyacetone or hydroxyethanal in bio-oil can oligomerize and polymerize. Aldol condensation reactions accelerate bio-oil aging [7]. Table 8 summarizes the effect of the presence of hydroxyacetone upon 1-octene reactions with phenol/water over $Cs_{2.5}/K10$ and SSA catalysts. A good phenol conversion (64.0%) was obtained over SSA. $Cs_{2.5}/K10$ was deactivated by hydroxyacetone based on the greatly reduced phenol conversion (2.2%). No hydroxyacetone was detected after reaction. However, small amounts of 3-methyl-2-hydroxycyclopent- 2-enone were detected. To further study this issue, neat 1-hydroxyacetone was heated at 100 °C for 1h over 30% $Cs_{2.5}/K10$ and the products were identified by GC-MS (Table 9) [22]. Propionic acid (43.7%), hydroxyacetone dimers (24.8%) and 2-hydroxy-3-methylcyclopent-2-enone (9.7%) formed, together with about 1.8% of an unknown species. Three carbon α-hydroxycarbonyl species such as 1-hydroxyacetone (acetol) can undergo enolization and dimerization to yield the structures illustrated in Scheme 5. The keto form of 1-hydroxyacetone exists in equilibrium with its enol, enediol and aldehydo forms. In both neat and concentrated solutions, 1-hydroxyacetone can dimerize generating the cyclic structures (a) and (b). Decomposition of (b) produces propionic acid. Meanwhile, intermolecular aldol condensation reactions of 1-hydroxyacetone and subsequent serial of dehydration reactions occurred generating 3-methyl-2-hydroxycyclopent-2-enone (c) and its isomers. As a consequence of dimerization and aldol condensation, the 1-hydroxyacetone monomers are expected to diminish with time during upgrading.

TABLE 7: Yields of butyl octyl ethers and phenol conversions in acid-catalyzed 1-octene reactions with phenol/water/1-butanol over different catalysts at 120 °C[a].

Catalyst	Phenol Conversion (%)	[b]Yield (%) Butyl octyl ethers
Amberlyst 15	40.7	0.17
$Cs_{2.5}/K10$	6.9	nd
Dowex50WX2	40.6	0.69
SSA	69.2	0.24

[a] Reaction conditions: (solid catalyst, 0.15 g; time: 3 h; 1-octene: phenol: water: 1-butanol (mmol) = 10:10:11.6:3.4; [b] GC area% of involved compounds versus the sum of the GC area% of all products remained after the reaction.

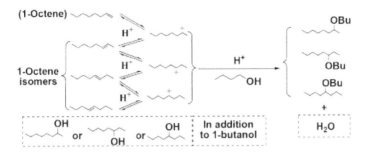

SCHEME 4: Acid-catalyzed reactions of 1-butanol as a model alcohol with 1-octene.

TABLE 8: Phenol conversions and yields of new products derived from the added reagent (2-hydroxymethylfuran or hydroxyacetone or D-glucose) in 1-octene reactions with water/phenol[a].

Added Reagent	Catalyst	Phenol conversion (%)	[d]Products yield (%)
Hydroxyacetone	SSA	64.0	Methyl cyclopentenolone (0.12)
	$Cs_{2.5}$/K10	2.2	Methyl cyclopentenolone, (0.21)
2-Hydroxymeth-ylfuran	SSA	65.0	UN[b] (0.18)
	$Cs_{2.5}$/K10	4.5	UN[b] (0.87)
D-glucose	SSA	64.7	Anhydrosugar[c] (0.14)
	$Cs_{2.5}$/K10	3.5	Anhydrosugar[c] (0.20)
	A15	40.5	octyl formates, 0.28
	DX2	41.3	octyl formates, 0.36

[a]Reaction conditions: (120 °C, 3h, catalyst, 0.15 g; 1-octene (10 mmol); phenol (10 mmol); water (10mmol); hydroxyacetone (1.4 mmol) or 2-hydroxymethylfuran (1.0 mmol) or D-glucose (0.6mmol)); [b]Unkown compound; [c]Anhydrosugar: 1,6-Anhydro-3,4-dideoxy-β-D-manno-hexapyranose; [d] GC area% of involved compounds versus the sum of the GC area% of all products remained after the reaction.

Furan derivatives such as 2-hydroxymethylfuran and hydroxymethyl-furfural present in bio-oil can polymerize easily and give tar in the presence of acid [26]. Table 8 shows the effect of 2-hydroxymethylfuran addition on the 1-octene reaction with phenol/water over $Cs_{2.5}/K10$ and SSA catalysts. SSA shows a good catalytic activity giving a phenol conversion of 64.7%. However, the $Cs_{2.5}/K10$ surfaces coked seriously, which greatly decreased its catalytic activity and reduced phenol conversion (4.5%). Polymerization of 2-hydroxymethylfuran occurred under acidic conditions forming coke or tar on the catalysts [27]. This has been further confirmed by the reaction products identified from GC-MS analysis (Table 8) of the reaction where neat 2-hydroxymethylfuran (2 g) was heated at 100 °C for or 1h over 30% $Cs_{2.5}/K10$. In addition to the unreacted 2-hydroxymethyl-furan, difurfuryl ether (9.4%), 5-furfuryl-furfuryl alcohol (4.9%) and difu-ran-2-ylmethane (4.8%), together with about 2.6% other compounds were detected in the products [22]. Rapid polymerization of 2-hydroxymeth-ylfuran occurred at 120 °C via acid-catalyzed electrophilic condensation with the accompanying loss of formaldehyde [18], generating difurfuryl ether, 5-furfuryl-furfuryl alcohol, difuran-2-ylmethane, etc. (Scheme 6).

TABLE 9: Products formed from hydroxyacetone or 2-hydroxymethylfuran on heating at 100 °C over 30% Cs2.5/K10 for 1 h[a].

2-Hydroxymethylfuran		Hydroxyacetone	
Compounds	Peak area (%)	Compounds	Peak area (%)
2-hydroxymethylfuran	78.3	Propionic acid	63.4
Difurfuryl ether	9.4	Hydroxyacetone dimers	24.8
5-Furfuryl-furfuryl alcohol	4.9	Methyl cyclopentenolone	9.7
Difuran-2-ylmethane	4.8	Unkown species	1.8
Others	2.6		

[a]*Products were analyzed by GC/MS. Each product was identified by excellent matches of their MS fragmentation patterns.*

A large number of compounds with hydroxyl groups, particularly, an-hydro monosaccharides such as levoglucosan, were derived from pyrolysis of cellulose and hemicellulose components of the pine wood feed during bio-oil production [5]. Anhydro monosaccharides readily hydrate back to monosaccharides when heated with acid and water. Therefore, the effects of D-glucose addition on 1-octene reactions with phenol/water over $Cs_{2.5}$/ K10, DX2, A15 and SSA catalysts were studied at 120 °C for 3 h (Table 8). SSA gives a higher activity than DX2, A15 and $Cs_{2.5}$/K10. Trace amounts of octyl formates were formed over A15 and DX2 by esterification of formic acid by octanols. The formic acid was generated by the known acid-catalyzed conversions pathways of D-glucose (Scheme 7) [2,27]. 5-Hydroxymethylfurfural (HMF) was formed by D-glucose dehydration. Then, subsequent HMF hydration to its hemiacetal occurred in acidic media, followed by rehydration, ring-opening, loss of both water and formic acid to form levulinic acid. Also, small amounts of 1,6-anhydro-3,4-dideoxy-β-D-mannohexapyranose, generated by dehydration of D-glucose, were detected over SSA and $Cs_{2.5}$/K10.

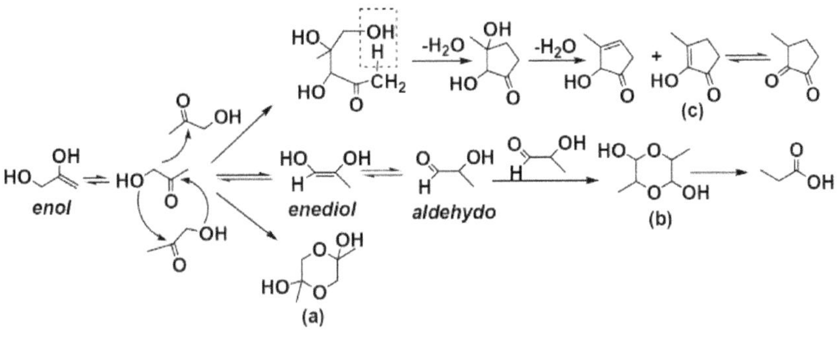

SCHEME 5: Acid-catalyzed reactions of hydroxyacetone.

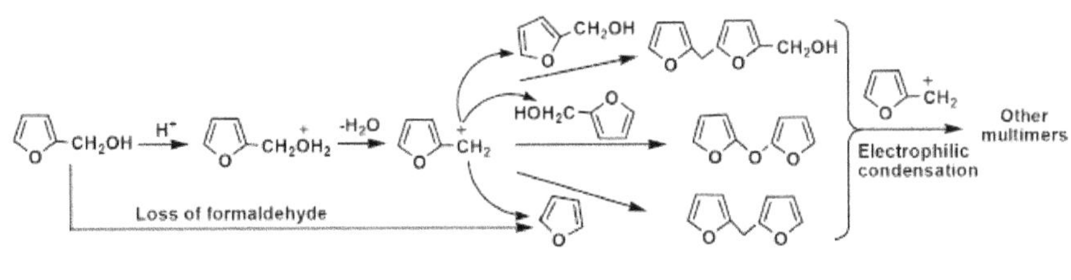

SCHEME 6: Acid-catalyzed reactions of 2-hydroxymethylfuran.

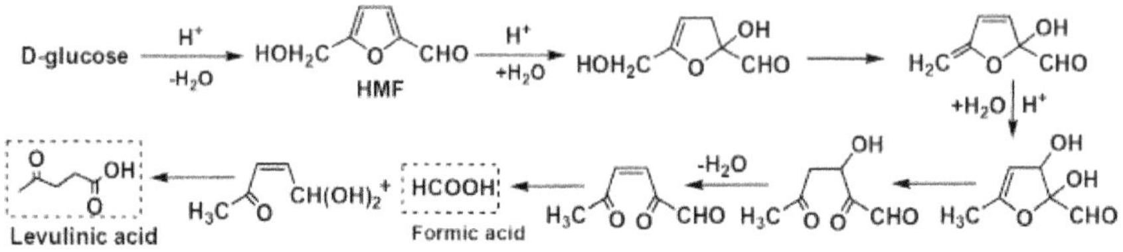

SCHEME 7: Acid-catalyzed reactions of D-glucose as model monosaccharide [27].

In addition to the reactions discussed above occurring between 1-octene with model bio-oil components, some additional olefin reactions took place. Table 10 shows the product distributions of individual olefin reactions conducted using equimolar amounts of olefin (1,7-octadiene, cyclohexene, 1-octene, or 2,4,4-trimethylpentene)/phenol/water over SSA at 120 °C for 3 h. Skeletal isomerization reactions of all olefins occurred except with cyclohexene. Intramolecular diene cyclizations occurred for 1,7-octadiene, individually.

No oligomerization or cracking of either 1-octene or 1,7-octadiene occurred. However, 1-octene isomers were detected in the reactions where neat phenol and 1-octene operated at 100 °C over $Cs_{2.5}$/K10 or A15 catalysts. 2,4,4-Trimethylpentene oligomerized to C16 olefins and cracked into C4 olefins readily at these conditions. Cracking of 2,4,4-trimethylpentene mainly produced isobutene. Cyclohexene's dimer 1-cyclohexyl-1-cyclohexene was found. Reoligomerization of isobutene gave C8, C12, and C16 olefins. Hydration reactions to form alcohols occurred for all olefins, followed by etherification of these resulting alcohols generating ethers and water (Scheme 8). Thus, the choice of olefin structure will play a role in the product distribution observed, but all the olefins help drive the upgrading process to remove water and promote esterification, acetal formation, generate ethers and both O- and C-alkylate phenols.

TABLE 10: Product distributions and phenol conversions in addition reactions by different olefins in phenol/water at 120 °C in 3h over the SSA catalyst[a].

Olefins	PC (%)[b]	Product distributions (GC area %)[c]
1,7-Octadiene	52.4	Skeletal isomers (4.5%): 1,6-octadiene, 3,5-octadiene, 2-methyl-1,5-heptadiene, 2,4-dimethyl-1,5-hexadiene, 3-methyl-1,5-heptadiene, etc.
		Cycloolefins (14.2%): bicyclo[4.1.0]heptane, 2-methylbicyclo[2.2.1]heptane, 4-ethyl-1-cyclohexene,cyclooctene,1-ethyl-2-methylcyclopen-tene, 1,6-dimethyl-cyclohexene,ethylidenecycloh exane, 1,2-dimethyl-1-cyclohexene, etc.
		Alcohols and ethers (8.8%): 1,7-octanediol, 4-ethylcyclohexanol, 1-ethylcyclohexanol,
		2-propyl-tetrahydropyran, 2,5-diethyltetrahydrofur-an, 2-propyltet-rahydropyran, 2-butyl-3-ethyloxirane, etc.
		O-Alkylates (33.4%); Mono-C-alkylates (4.1%); Di-C-alkylates (10.1%)
Cyclohexene	62.0	Oligomers (0.2%):1-cyclohexyl-1-cyclohexene, etc.
		Alcohols and ethers(1.4%):cyclohexanol, di-cyclohexyl ether
		O-Alkylates (2.9%): cyclohexyl phenyl ether
		Mono-C-alkylates (33.4%): o-cyclohexylphenol, p-cyclohexylphenol
		Di-C-alkylates (26.8%); Tri-C-alkylates (8.3%):2,4,6-tricyclohexyl-phenol
1-Octene	74.5	Skeletal isomers (12.1%):2-octene, 3-octene, 4-ocene, 1-octene.
		Alcohols and ethers (5.1%):2-octanol, 3-octanol dioctyl ethers, etc.
		O-Alkylates (48.6%):2-Octyl phenyl ether, 3-Octyl phenyl ether, etc.
		Mono-C-alkylates (14.5%): o-octylphenol, p-octylphenols, etc.
		Di-C-alkylates (11.5%): 2,4-di-octylphenols, 2,6-dioctylphenols,
2,4,4-Trimethyl-pentene	89.9	Skeletal isomers (4.7%): 2,4,4-trimethyl-1-pentene, 3,4,4-trimethyl-2-pentene, etc.
		Fragments and oligomers (3.8%): 4,4-dimethyl-2-neopentyl-1-pentene, 2,2,4,6,6-pentamethyl-3-heptene, 2,2,4,4,6,8,8-heptamethyl-nonane, 2,4,4,6,6,8,8-heptamethyl-1-nonene, 2,2,4,6,6- pentamethyl-3-heptene,etc.
		Alcohols and ethers (1.8%): 2,5,5-trimethyl-2-pentanol
		Mono-C-alkylates (64.2%): p-(1,1,3,3-tetramethylbutyl)phenol, o-t-butylphenol, p-t-butylphenol, etc.
		Di-C-alkylates (12.1%): 2,4-di-t-butylphenol, 2,5-di-t-butylphenol, 2,6-di-t-butylphenol, 2-t-butyl-4-(1,1,3,3-tetramethylbutyl)phenol, etc.
		Tri-C-alkylates (6.7%): 2,4,6-tri-t-butylphenol

[a]Reaction conditions: (SSA, 0.15 g; olefin: phenol: water (mmol) = 10:10:10; [b]Phenol conversions; [c]GC area% of involved compounds versus the sum of the GC area% of all products remained after the reaction.

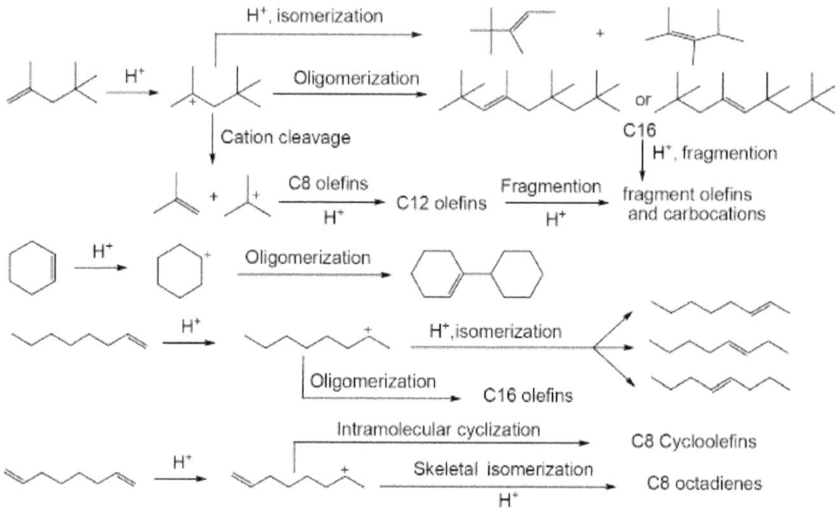

SCHEME 8: Acid-catalyzed reactions of olefin reagents.

TABLE 11: Product distributions and 1-butanol conversions in acid-catalyzed 1-octene/1-butanol reactions with phenol/water at 120 °C[a].

Catalyst	1-Buta-nol Conv. (%)	Selectivity (%)				Yield (%)[b]		
		Mono-O-octylates	Mono-C-octylates	Di-octy-lates	Octa-nols	Dioctyl ethers	Dibutyl ether	O-Bu-tylate
A15	60.9	84.2	14.5	1.2	6.3	2.0	11.9	0.14
DX2	58.7	84.8	14.1	1.0	6.9	2.1	11.9	0.61
SSA	74.8	82.3	13.8	3.9	4.8	1.9	12.4	0.1
Cs$_{2.5}$/K10	8.2	53.4	46.6	0	0.1	Nd	0.1	Nd

[a]Reaction conditions: (solid catalyst, 0.15 g; time: 3 h; 1-octene: phenol: water: 1-butanol (mmol) = 10:10:11.6:3.4; b GC area% of involved compounds versus the sum of the GC area% of all products remained after the reaction.

In addition, intermolecular dehydration of 1-butanol occurred producing dibutyl ether with both Dowex50WX2 and SSA (Table 11). Also, traces of butyl phenyl ether were formed. Small amounts of t-butyl phenols

(0.9%) were detected when 2-butanol was used as replacement of 1-butanol over SSA. Under acid catalyzed conditions, 2-butanol was protonated, and then dehydrated generating secondary carbocations. Some isomerization to tertiary carbocations must then occur. Phenol added to these C4 cations generating O-t-butylated phenol, followed by isomerization of the t-butyl phenyl ether to the thermodynamic C-t-butylated phenol products along with some bis-alkylated phenols formation (Scheme 9). Also, both acetic acid esterification and acetaldehyde acetalation reactions with 1-butanol occurred, generating butyl acetate and 1,1-dibutoxyethane (acetal), respectively, when acetic acid and acetaldehyde were present.

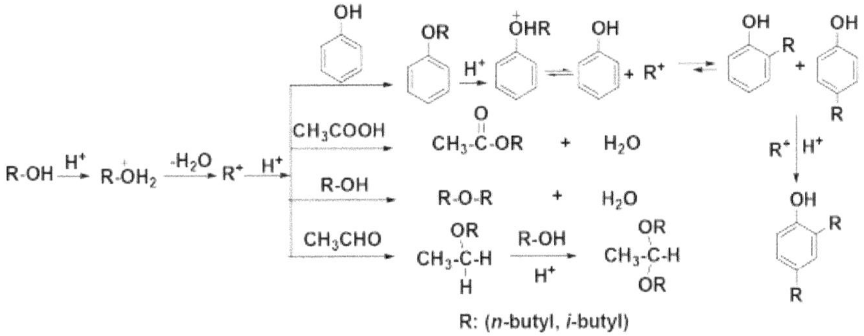

SCHEME 9: Acid-catalyzed reactions of model bio-oil components with alcohol.

Table 12 summarizes the product compositions of the model bio-oil (phenol/water/acetic acid/acetaldehyde/hydroxyacetone/D-glucose/2-hydroxymethylfuran mixtures) reactions with 1-octene/1-butanol over all five acid catalysts at 120 °C for 3 h [22].

In addition to alkylated phenols (both O-octylated and C-octylated), octanols, dibutyl ether and dioctyl ether, butyl acetate and various octyl acetates, 1-octene oligomers and isomers mentioned above, 1,1-dibutoxyethane and butyl levulinate were formed over all the catalysts. 1,1-Dibutoxyethane was formed by acetal formation between acetaldehyde and 1-butanol (Scheme 9). Levulinic acid reacted with 1-butanol forming butyl levulinate. Levulinic acid was derived from the acid-catalyzed conversions of both D-glucose (Scheme 7) and 2-hydroxymethylfuran (Scheme 10), which are both known pathways. Levulinic acid has been obtained by dehydration of hexoses to 5-hydroxymethylfurfural (HMF) and its subsequent hydration in acidic media [2,28]. Acid-catalyzed conversion of 2-hydroxymethylfuran to levulinic acid in aqueous solutions also has been recently reported [29].

TABLE 12: Product compositions of model bio-oil reactions with 1-octene/1-butanol over $Cs_{2.5}$/K10, A15, A36, DX2 and SSA catalysts at 120 °C in 3 h[a].

Products	Peak area (%)				
	$Cs_{2.5}$/K10	A36	A15	DX2	SSA
Unreacted					
1-octene	48.4	54.1	26.4	20.3	4.0
1-butanol	1.0	0.2	0.1	0.3	0.2
phenol	40.4	21.9	15.7	15.3	9.6
1-Dodecane (Internal standard)	1.4	1.4	1.5	1.8	3.0
In common					
1-Octene isomers	1.0	8.5	31.1	25.3	28.9
Phenol octylates	0.8	2.9	12.4	18.8	35.3
Octanols	0.1	2.3	2.6	3.8	1.5
Dioctyl ethers	0.1	Nd	0.1	0.3	1.3
[b]1-Octene oligmers and their hydrates	3.8	4.2	4.7	5.0	6.1
Octyl acetates	0.8	0.9	1.6	6.7	4.8
1,1-Dibutoxyethane	0.3	0.1	0.1	0.1	0.1
Dibutyl ether	1.4	0.1	0.3	0.4	2.7
Butyl acetate	0.3	2.9	3.0	1.7	2.3
Butyl levulinate	0.1	0.6	0.3	0.2	0.4
Independent					
2-Hydroxy-3-methylcyclopent-2-enone	0.1	Nd	0.1	0.1	Nd
2-(2-Furylmethyl)furan	0.1	Nd	Nd	Nd	Nd

[a]Reaction conditions: (catalyst, 0.15 g; 1-octene: 1-butanol: phenol: water: acetic acid: acetaldehyde: hydroxyacetone: D-glucose: 2-hydroxymethylfuran (g) = 1.35: 0.15: 0.94: 0.15: 0.15: 0.12: 0.12: 0.15: 0.15; time: 3 h; nd: not detected); these results were first noted in the preliminary communication of this work found in reference 23; [b]C8, C16 and C24 olefins and their hydrates.

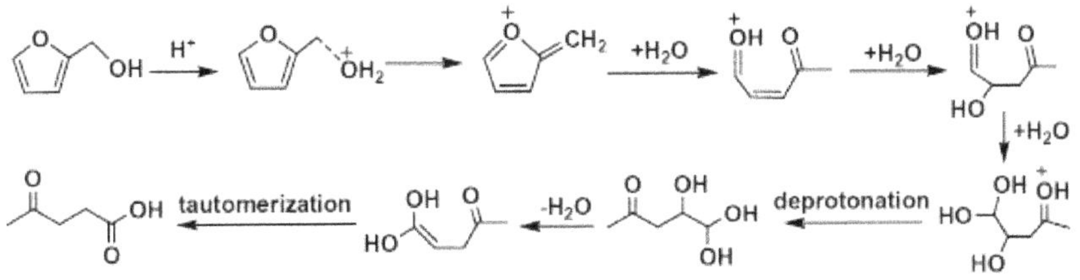

SCHEME 10: Acid-catalyzed conversion of 2-hydroxymethylfuran to levulinic acid in aqueous solution [29].

Butyl levulinate shows the extent of levulinic acid formation. Table 13 demonstrates that it originates from both D-glucose and 2-hydroxymethylfuran. Increasing butyl levulinate formation was observed when the amount of 2-hydroxymethylfuran and D-glucose were increased in the model bio-oil [22]. Trace amounts of 3-methyl-2-hydroxycyclopent-2-enone were generated from hydroxyacetone over $Cs_{2.5}/K10$, A15 and DX2 catalysts. Among the new products formed, 1-octene oligomers were the most abundant components over $Cs_{2.5}/K10$ or A36. However, phenol alkylates became the most abundant products for the other three catalysts. Trace amount of 2-(2-furylmethyl) furan, formed by furfuryl alcohol polymerization, was detected over the $Cs_{2.5}/K10$ catalyst. No 2-(2-furylmethyl) furan was detected over other catalysts, because all the reactive furan products were consumed by these more acidic catalysts.

5.2.3 PROPOSED REACTION PATHWAYS FOR MODEL BIO-OIL COMPONENTS

A complicated but clear reaction pathway is postulated and shown in Figure 2 based on the detailed product analyses and discussions mentioned above. Under acid-catalyzed conditions, olefin protonation and subsequent proton loss and reprotonation steps generated the isomerized olefins and their cation intermediates. Simultaneously, a series of competing reactions occur, where bio-oil's components (water, carboxylic acids, phenols and alcohols) and the added olefins add to these cations. This leads to hydration, esterification, O-alkylation, etherification and oligomerization which forms alcohols, esters, phenol O-alkylates, ethers and olefin oligomers, respectively. Moreover, diene intermolecular cyclizations and branched olefin cracking into small fragments as well as reoligomerization of these small fragments occurred. Similarly, under acid catalyzed conditions, protonation of alcohols and subsequent dehydration of these protonated products occurred generating carbocations. Meanwhile, additional competing reactions among carboxylic acids, aldehydes, alcohols, phenols and levulinic acid with these carbocations occurred, generating esters, acetals, ethers, phenol O-alkylates and alkyl levulinates, respectively. O-Alkylated phenols isomerized to the thermodynamic C-alkylated phenol via a Friedel Crafts mechanism. Further addition of carbocations to mono-alkylated phenols generated bis-alkylated phenols.

TABLE 13. Butyl levulinate yields obtained from reactions of 1-octene/1-butanol with model bio-oils containing different amounts of D-glucose, 2-hydroxymethylfuran and 1-butanol at 120 °C in 3h over Dowex50WX2[a].

Addition amounts (g)			Butyl levulinate yields (Area %)
2-Hydroxymethylfuran	D-glucose	1-Butanol	
0.15	0.15	0.15	0.22
0.15	0.30	0.30	0.34
0.15	0.30	0.30	0.98
0.30	0.15	0.30	1.39
0.30	0.30	0.60	2.45

[a]*Reaction conditions: (catalyst, 0.15 g; 1-octene: 1-butanol: phenol: water: acetic acid: acetaldehyde: hydroxyacetone: D-glucose: 2-hydroxymethylfuran (g) = 1.35: (0.15~0.6): 0.94: 0.15: 0.15: 0.12: 0.12: (0.15~0.3): (0.15~0.3); these results were first noted in the preliminary communication of this work found in Reference [22].*

In addition to reactions between bio-oil components and olefin/alcohol reagents, the added alcohol adds across olefins to give intermolecular etherification. Self-etherification of the added alcohol reagent also occurs. These reactions occur simultaneously, generating corresponding ethers. Acid-catalyzed dehydration of D-glucose to levulinic acid [28] occurred. First, dehydration of D-glucose gives 5-hydroxymethylfurfural (HMF). Then, HMF hydration to its hemiacetal occurred followed by sequential rehydration, ring-opening, loss of both water and formic acid generating levulinic acid. Also, acid-catalyzed 2-hydroxymethylfuran rehydration and subsequent ring opening, dehydration and tautomerization formed levulinic acid [29]. Levulinic acid, in turn, is converted to alkyl levulinates by alcohols. Independently, polymerization of 2-hydroxymethylfuran via electrophilic aromatic substitution proceeded jointly with loss of formaldehyde to form oligomeric products. Simultaneous dimerization and isomerization of hydroxyacetone occurred, forming cyclic hydroxyacetone dimers and propionic acid along with some 2-hydroxy-3-methylcyclopent-2-enone. This latter product was likely formed by aldol condensation of hydroxyacetone to an open hydroxyacetone dimer and its subsequent dehydration and cyclodehydration reactions.

Clearly, bio-oil upgrading by simultaneous reactions with olefin/alcohol over solid acids is complex, involving many simultaneous equilibria and competing reactions. However, the key reason for the success of this upgrading process is the role of acid-catalyzed olefin hydration. Olefin hydration removes water. As water concentration drops, esterification and acetal formation equilibria shift toward ester and acetal products.

5.3 EXPERIMENTAL SECTION

All chemicals were purchased from Sigma Aldrich (St. Louis, MO, USA), and used without further purification unless otherwise noted.

5.3.1 CATALYST PREPARATION

The silica gel, Kieselgel 40 (4 nm mean pore diameter, 590 $m^2 \cdot g^{-1}$), was dried at 120 °C for 3 h in air prior to its use. The following SSA catalyst was prepared by a well-developed procedure [23,30]. A 250 mL suction flask, equipped with a constant pressure dropping funnel containing 5.83

g chlorosulfonic acid and a gas inlet tube for releasing HCl gas, was charged with 10.0 g Kieselgel 40 silica gel and 50 mL CH_2Cl_2. Chlorosulfonic acid was added dropwise over 30 min while stirring at room temperature. HCl gas was immediately evolved and absorbed into water. The mixture was then stirred for another 30 min. Next, CH_2Cl_2 was removed by rotary evaporation (50 °C, 20 min). A white solid (SSA, yield, 98%) was obtained and stored in a desiccator until use. K10 clay supported $Cs_{2.5}H_{0.5}PW_{12}O_{40}$ catalyst (hereafter designated $Cs_{2.5}$/K10) was prepared by a well-developed route [31]. Typical Fourier Transform Infrared (FTIR) spectra of these two catalysts were recorded on a Thermo Scientific Nicolet 6700 spectrophotometer. Surface area, total pore volume and pore diameter of the catalysts were determined by N_2 adsorption at 77 K using a Quantachrome Nova 2000 instrument after evacuating at 393 K for 3 h under nitrogen atmospheric. The total amount of acidity (H^+) was measured by titration of catalyst samples in water with standardized sodium hydroxide (0.495 M).

5.3.2 CATALYTIC REACTIONS

All reactions were carried out in glass pressure reaction vessels equipped with a magnetic stirrer. The temperature, controlled using an external oil bath, was raised to the desired value (100 °C or 120 °C) and held for the desired time (1 h or 3 h) with vigorous stirring. In a typical reaction, SSA (0.15 g), 1-octene (1.35 g), 1-butanol (0.15 g), phenol (0.94 g), water (0.15 g), acetic acid (0.15 g), acetaldehyde (0.12 g), hydroxyacetone (0.12 g), D-glucose (0.15 g), 2-hydroxymethylfuran (0.15 g) and the internal standard (99.9% 1-dodecane, 0.02 g) were charged in that order. Catalysts studied included SSA, $Cs_{2.5}$/K10, A15, A36 and DX2. After reaction (typically, 3 h), all products were diluted in methanol and identified by analysis on a Shimadzu QP2010S gas chromatograph equipped with a mass selective detector (GC-MS) using helium as the carrier gas. A SHRXI-5MS (30 m × 0.25 mm I.D. × 0.25 µm film) capillary column was used with a 50:1 split ratio and a solvent cut time of 3 min. The temperature program, started at 30 °C (5 min), was ramped from 30 to 300 °C at 10 °C/min and held at

300 °C for 8 min. An auto-sampler and the same analysis method were used for all product analyses. MS identification of the products was based on molecular mass, fragmentation patterns and by matching the spectra with a digital compound library. The percent phenol conversion to other products in the upgrading reactions was determined by the change in peak area versus that of the 1-dodecane internal standard.

5.4 CONCLUSIONS

Liquid phase supported acid-catalyzed olefin/alcohol reactions with model bio-oils indicate that silica sulfuric acid is an improved catalyst with greater hydrothermal stability and catalytic activity over $Cs_{2.5}$/K10 and other resin sulfonic acids. Development and demonstration of this improved catalyst meets one goal of this study. $Cs_{2.5}$/K10 lost most of its catalytic activity, poisoned by the coke formation from hydroxyacetone, 2-hydroxymethylfuran and D-glucose. Decomposition of resin-bound sulfonic acids occurred.

The use of different olefins and alcohols leads to different product selectivities. This study has demonstrated many of the competing reaction pathways which occur in bio-oil upgrading by acid-catalyzed alcohol/olefin treatment in much greater detail than all previous work, thereby accomplishing a second major goal of this work. Upgrading bio-oil via simultaneous reactions with olefin/alcohol under acid-catalyzed conditions was complex, involving many simultaneous equilibria and competing reactions. These reactions mainly include phenol alkylation, olefin hydration, esterification, etherification, acetal formation, olefin isomerization and oligomerization, cracking and reoligomerization of tertiary cation centers from protonated olefins and their fragments, hydroxyacetone dimerization (including cyclization) and intermolecular aldol condensation. Also, levulinic acid formation both from sequential dehydration, ring contractions, hydrations and ring opening of monosaccharides, and from sequential rehydration, ring opening, dehydration and tautomerization of 2-hydroxymethylfuran occurred. Synergistic interactions among reactants and products were determined.

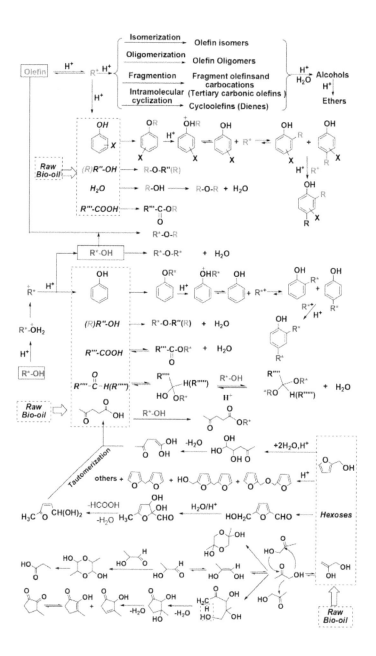

FIGURE 2: Reaction pathways of the model bio-oil components during upgrading with olefin/alcohol over solid acid catalysts.

Water removal by acid-catalyzed olefin hydration is the key reason for the success of this upgrading process. As water concentration drops, esterification and acetal formation equilibria shift toward ester and acetal products. In turn, the formed esters and acetals as well as the added alcohol help reduce the phase separation present between hydrophilic bio-oil and hydrophobic olefin. All of this occurs while maintaining all the caloric value of both the raw bio-oil and the alcohol and olefin reagents. This work also provides further insight into the complexity of this bio-oil upgrading approach.

REFERENCES

1.   Ragauskas, A.J.; Williams, C.K.; Davison, B.H.; Britovsek, G.; Cairney, J.; Eckert, C.A.; Frederick, W.J., Jr.; Hallett, J.P.; Leak, D.J.; Liotta, C.L.; et al. The path forward for biofuels and biomaterials. Science 2006, 311, 484–489.

2.   Huber, G.W.; Iborra, S.; Corma, A. Synthesis of transportation fuels from biomass: Chemistry, catalysts, and engineering. Chem. Rev. 2006, 106, 4044–4098.

3.   Amidon, T.E.; Wood, C.D.; Shupe, A.M.; Wang, Y.; Graves, M.; Liu, S.J. Biorefinery: Conversion of woody biomass to chemicals, energy and materials. J. Biobased Mater. Bioenergy 2008, 2, 100–120.

4.   Butt, D. Thermochemical Processing of Agroforestry Biomass for Furans, Phenols, Cellulose and Essential Oils; Final Report for RIRDC Project PN99.2006; RIRDC Publication (No.06/121): Melbourne, Australia, 2006.

5.   Mohan, D.; Pittman, C.U., Jr.; Steele, P.H. Pyrolysis of wood/biomass for bio-oil: A critical review. Energy Fuels 2006, 20, 848–889.

6.   Czernik, S.; Bridgwater, A.V. Overview of applications of biomass fast pyrolysis oil. Energy Fuels 2004, 18, 590–598.

7.   Diebold, J.P. A review of the chemical and physical mechanisms of the storage stability of fast pyrolysis bio-oils. In Fast Pyrolysis of Biomass: A Handbook; CPL Scientific Publishing: Newbury, UK, 2002; Volume 2, pp. 243–292.

8.   Elliot, D.C.; Hart, T.R.; Neuenschwander, G.G.; Rotness, L.J.; Zacher, A.H. Catalytic hydroprocessing of biomass fast pyrolysis bio-oil to produce hydrocarbon products. Environ. Progr. Sustain. Energy 2009, 28, 441–449.

9.   Ben, H.; Mu, W.; Deng, Y.; Ragauskas, A.J. Production of renewable gasoline from aqueous phase hydrogenation of lignin pyrolysis oil. Fuel 2013, 103, 1148–1153.

10. Venderbosch, R.H.; Prins, W. Fast pyrolysis technology development. Biofuels Bioprod. Biorefin. 2010, 4, 178–208.

11. Mortensen, P.M.; Grunwaldt, J.D.; Jensen, P.A.; Knudsen, K.G.; Jensen, A.D. A review of catalytic upgrading of bio-oil to engine fuels. Appl. Catal. A 2011, 407, 1–19.

12. Bulushev, D.A.; Ross, J.R.H. Catalysis for conversion of biomass to fuels via pyrolysis and gasification: A review. Catal. Today 2011, 171, 1–13.
13. Bridgwater, A.V. Review of fast pyrolysis of biomass and product upgrading. Biomass Bioenergy 2012, 38, 68–94.
14. Zhang, H.; Xiao, R.; Wang, D.; He, G.; Shao, S.; Zhang, J.; Zhong, Z. Biomass fast pyrolysis in a fluidized bed reactor under N2, CO2, CO., CH4 and H2 atmospheres. Bioresour. Technol. 2011, 102, 4258–4264.
15. Zabeti, M.; Nguyen, T.S.; Lefferts, L.; Heeres, H.J.; Seshan, K. In situ catalytic pyrolysis of lignocellulose using alkali-modified amorphous silica alumina. Bioresour. Technol. 2012, 118, 374–381.
16. Wan, Y.Q.; Chen, P.; Zhang, B.; Yang, C.; Liu, Y.; Lin, X.; Ruan, R. Microwave-assisted pyrolysis of biomass: Catalysts to improve product selectivity. J. Anal. Appl. Pyrolysis 2009, 86, 161–167.
17. Zhang, Z.J.; Wang, Q.W.; Yang, X.L.; Chatterjee, S.; Pittman, C.U., Jr. Sulfonic acid resin-catalyzed addition of phenols, carboxylic acids, and water to olefins: Model reactions for catalytic upgrading of bio-oil. Bioresour. Technol. 2010, 101, 3685–3695.
18. Yang, X.L.; Chatterjee, S.; Zhang, Z.J.; Zhu, X.F.; Pittman, C.U., Jr. Reactions of phenol, water, acetic acid, methanol, and 2-hydroxymethylfuran with olefins as models for bio-oil upgrading. Ind. Eng. Chem. Res. 2010, 49, 2003–2013.
19. Zhang, Z.J.; Wang, Q.W.; Tripathi, P.; Pittman, C.U., Jr. Catalytic upgrading of bio-oil using 1-octene and 1-butanol over sulfonic acid resin catalysts. Green Chem. 2011, 13, 940–949.
20. Cheng, C.L.; Chien, L.J.; Lee, W.J.; Che, P.Y.; Chen, B.Y.; Chang, J.S. High yield bio-butanol production by solvent-producing bacterial microflora. Bioresour. Technol. 2012, 113, 58–64.
21. DuPont/BP. DuPont and BP Announce Partnership to Develop Advanced Biofuel; DuPont Press Release: Wilmington, NC, USA, 2006. Available online: http://multivu.prnewswire.com/mnr/dupont/24656/ (accessed on 4 March 2013).
22. Zhang, Z.J.; Sui, S.J.; Tan, S.; Wang, Q.W.; Pittman, C.U., Jr. Catalytic conversion of bio-oil to oxygen-containing fuels by simultaneous reactions with 1-butanol and 1-octene over solid acids: Model compound studies and reaction pathways. Bioresour. Technol. 2012, 130, 789–792.
23. Minakata, S.; Komatsu, M. Organic reactions on silica in water. Chem. Rev. 2009, 109, 711–724.
24. Meng, F.B.; Zhang, B.Y.; Liu, L.M.; Zang, B.L. Liquid-crystalline elastomers produced by chemical crosslinking agents containing sulfonic acid groups. Polymer 2003, 44, 3935–3943.
25. Marcelo, E.; Domine, C.; Veen, V.; Schuurman, Y.; Mirodatos, C. Coprocessing of oxygenated biomass compounds and hydrocarbons for the production of sustainable fuel. ChemSusChem 2008, 1, 179–181.
26. Corma, A.; Hube, G.W.; Sauvanaud, L.; O'Connor, P. Biomass to chemicals: Catalytic conversion of glycerol/water mixtures into acrolein, reaction network. J. Catal. 2008, 257, 163–171.
27. Kim, T.; Assary, R.S.; Marshall, C.L. Gosztola, D.J.; Curtiss, L.A.; Stair, P.C. Acid-catalyzed furfuryl alcohol polymerization: Characterizations of molecular structure and thermodynamic properties. ChemCatChem 2011, 3, 1451–1458.

28. Horvat, J.; Klaic, B.; Metelko, B.; Sunjic, V. Mechanism of levulinic acid formation. Tetrahedron. Lett. 1985, 26, 2111–2114.

29. González Maldonado, G.M.; Assary, R.S.; Dumesic, J.; Curtiss, L.A. Experimental and theoretical studies of the acid-catalyzed conversion of furfuryl alcohol to levulinic acid in aqueous solution. Energy Environ. Sci. 2012, 5, 6981–6989.

30. Shaabani, A.; Rahmati, A. Silica sulfuric acid as an efficient and recoverable catalyst for the synthesis of trisubstituted imidazoles. J. Mol. Catal. A 2006, 249, 246–248.

31. Kamalakar, G.; Komura, K.; Sugi, Y. Tungstophosphoric acid supported on MCM-41 mesoporous silicate: An efficient catalyst for the di-tert-butylation of cresols with tert-butanol in supercritical carbon dioxide. Appl. Catal. A Gen. 2006, 310, 155–163.

CHAPTER 6

Emission Abatement at Small-Scale Biomass Combustion Unit with High-Temperature Catalysts

R. BINDIG, S. BUTT, AND I. HARTMANN

6.1 INTRODUCTION

The use of biomass or bioenergy can be traced back to the beginning of human civilization when people started to burn wood for heating and cooking purposes. Ironically, after so many years have gone by, wood still remains the largest biomass resource in the world [1]. However, one major difference which has occurred over this period of time is the introduction of the concept "modern biomass" which states the usage of traditional biomass resources in highly efficient systems. This concept has been put into practice with more conviction and determination during the last decade, particularly in Europe, due to ever rising CO_2 levels in our environment. By now, it is an established fact that about 10-30% of total energy demand for hot water supply and domestic heating in European countries like, Austria, Germany, Sweden and Finland is provided by small scale biomass

Emission Abatement at Small-Scale Biomass Combustion Unit with High-Temperature Catalysts. © Bindig R, Butt S, and Hartmann I. Journal of Thermodynamics & Catalysis 4,125 (2013), doi: 10.4172/2157-7544.1000125. Licensed under Creative Commons Attribution License, http://creativecommons.org/licenses/by/4.0/.

combustion systems [2]. Moreover, it has been also concluded that despite the vast spread of technologically advanced small scale combustion devices in European countries (like countries mentioned above) during the recent years, still the old biomass combustion systems (stoves and boilers) occupy more consumers [3]. These conventional systems which are based on natural draft play a pivotal role in contributing to the high emission levels of particulate matter (PM), carbon monoxide (CO), organic gaseous compounds (OGC) and polycyclic aromatic hydrocarbons (PAH). These facts and figures have triggered an enormous understanding and awareness among the researchers as well as local population concerning harmful pollutants emitted by residential biomass combustion systems. For this reason inefficient small scale biomass combustion systems have been heavily criticized and demanded to be replaced by new efficient technologies.

Speaking of older and newer technologies, it has to be mentioned here that two types of technologies exist concerning small scale biomass combustion systems. The old biomass combustion systems are based on "up-burn" which is in a process of being rapidly replaced by "down-firing" systems (new technologies). As mentioned above, these older systems are a main source of PM1 (particles with diameter less than 1 μm) in European countries. It has been also concluded that such particles serve as a purpose of "support" onto which carbonaceous particles (organic compounds and soot) are deposited which are primarily responsible for the adverse health effects [4]. So in order to counter such an undesired release of pollutants, particularly from small scale biomass systems, a concept has been conceived according to which "down-firing" technology will be implemented in specially designed wood log stove in combination with catalytic treatment in order to abate harmful emissions to minimum possible values. It is noteworthy to mention here that the abatement of emissions through catalytic treatment from small scale biomass combustion systems has not been studied or implemented on a wide scale. So this novel concept of integrating catalytic components in different parts of the stove i.e. grate, walls of combustion chamber and the base will open more channels and schemes in order to accomplish the acceptable emission levels coming out of biomass combustion systems particularly, those used for residential purposes.

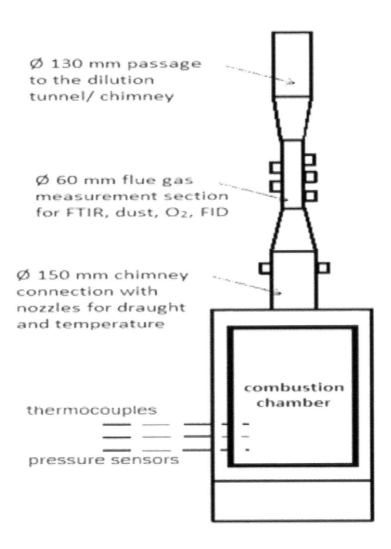

Ø 130 mm passage to the dilution tunnel/ chimney

Ø 60 mm flue gas measurement section for FTIR, dust, O₂, FID

Ø 150 mm chimney connection with nozzles for draught and temperature

combustion chamber

thermocouples

pressure sensors

FIGURE 1: Illustration of the test bench with a flue gas measurement section (hot) for the emission measurement.

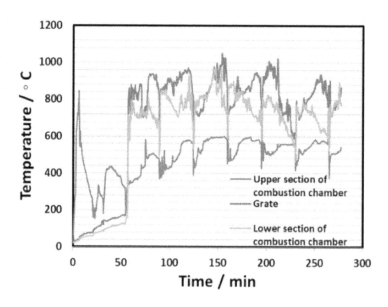

FIGURE 2: Time-dependent behavior of temperature during the reference experiment.

In the past, the process of catalysis has been strongly linked to chemical and refinery industries. However, recently the catalytic converters have been deployed and installed in automobiles, biomass fired boilers and power generation facilities in order to promote the environmentally friendly usage of technological devices. It has been estimated that the market of catalysis around the world worth around US$9 billion, out of which, one third is occupied by the environmental catalysis. So building on this ever growing trend of environmental catalysis, this article gives a further insight into the integration of catalytic components in a downdraft wood log stove to foresee the feasibility of this novel approach to resolve the problem of high emissions (e.g. carbon monoxide, volatile organic compounds, dust particles etc.) at small-scale furnaces for solid biomass.

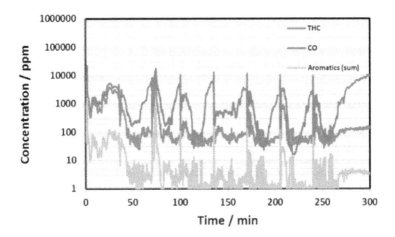

FIGURE 3: Time-dependent behavior of pollutants during the reference experiment.

6.2 MATERIALS AND METHODS

6.2.1 SETUP AND DESCRIPTION OF THE TESTED EQUIPMENT ALONG WITH MEASURING TECHNIQUES

A test bench has been developed (as shown in the Figure 1) in order to examine the emissions from a prototype downdraft stove. The test bench is designed in such a way that it can facilitate the analysis of dust composition.

For the determination of flue gas and combustion chamber temperature profiles, the thermocouples of Type K (manufactured by the company "Newport Electronics GmbH") have been used. For this purpose, a set of various thermocouples has been inserted into the grate, in the middle of upper and lower combustion chambers as well as in the walls of the lower combustion chamber. Moreover, the pressure conditions were recorded with the help of pressure sensors, inserted into the combustion chambers (upper and lower) as well as into the exhaust pipe. The measurement of static and dynamic pressures in the flue gas has been done with the aid of Prandtl tube produced by the company "Testo AG". The continuous transmission and data recording of Prandtl tube and pressure nozzles in the combustion chamber is carried out through data logging module provided by the company "Ahlborn". The data of the thermocouples have been recorded through a data logger of the company "National Instruments" along with the help of the software "Labview".

The emissions coming out of the stove are measured by means of a gas analyzer which consists of a Fourier Transform Infrared Spectrometer (FTIR, Manufacturer: Calcmet), a Flame Ionization Detector (FID, Manufacturer: Mess- & Analysentechnik GmbH, Typ: thermo-FID ES) and a paramagnetic oxygen analyzer (Manufacturer: M&C, Type: PMA 100). The infrared spectrum of FTIR can measure simultaneously organic as well as inorganic components. At the moment, about 44 different components can be recorded through FTIR.

The Volatile Organic Compounds (VOC) can be recorded by means of both FID and FTIR measuring devices. In case of VOC, the concentrations ranging under 50 mg/m^3 (at standard conditions i.e. =0°C, 1 atm) can be considered from the FID measuring device. On the other hand, the values above 50 mg/m^3 (at standard conditions i.e. =0°C, 1 atm) can be assumed from the FTIR measuring device. Following parameters can be measured simultaneously:

1. Oxygen O_2 (paramagnetic analyzer)
2. Carbon dioxide (FTIR)
3. Moisture in the flue gas i.e. H_2O (FTIR)
4. Carbon monoxide CO (FTIR)
5. Volatile organic compounds (VOC) as organic carbon (Org.-C) (FTIR and FID)
6. Nitrogen oxide as nitrogen dioxide equivalent (NO_{2equi}) (FTIR)
7. Sulphur dioxide SO_2 (FTIR)

8. Methane CH_4 (FTIR)
9. Organic compounds like, alkanes, alkenes, aromatics, aldehydes as well as ketones (FTIR)
10. Flue gas temperature, gas velocity and draft conditions.

The recording of the above mentioned parameters took place on continuous time basis except for the dust measurement. During the evaluation of the data, the average values of the pollutants were calculated for each dust sampling cycle whereas each cycle lasts for 30 minutes. With the aid of a chimney fan, a constant negative pressure of 12 Pa has been maintained in the chimney stack in order to achieve a fuel thermal output from 8 to 9 kW.

The gravimetric analysis of total amount of dust was done in accordance with VDI guidelines 2066-1, according to which a partial volume flow must be taken in isokinetic manner out of the main flue gas stream. In this process, the accompanied particles can be deposited on the already weighed plane filter. Since the filter housing is located outside the flue gas pipe, this sampling procedure is termed as "outstack process". The filter system was heated up with a heating jacket in order to prevent the falling down of temperature under saturation temperature of the flue gas. The temperature of the filter was maintained at 70°C so that the semi-volatile hydrocarbons could also be deposited on the filter. After the experiment, the deposited dust amount was determined gravimetrically and then can be specified by taking into consideration the measured partial volume and oxygen concentration. The plane filter was made of micro-glass fibers having a diameter of 45 mm.

6.2.2 CATALYST SYNTHESIS TECHNIQUES

In this work, two different synthesis routes have been developed in order to coat aluminium oxide foams with a mixed metal oxide active phase. These two respective techniques cannot be yet described in detail because of ongoing patent approval.

6.3 RESULTS AND DISCUSSION

6.3.1 REFERENCE EXPERIMENT

At first, a reference experiment was carried on the downdraft stove in order to determine the emissions, temperature profiles and pressure conditions during the operation of the stove in an unmodified state. This reference test is vital in the context of evaluating the effect of different modifications and changes in the stove which will be done in upcoming experiments. In Figure 2, the temperature profiles of different sections of the stove have been depicted. For every burning cycle, the stove was operated for the first 30 s in "up-draft" mode. After that, it was operated in downdraft (Twinfire mode) for the next 29.5 minutes. The average temperature in the grate was calculated to be around 750°C whereas, the temperature in the walls of the lower combustion chamber, where catalysts are planned to be installed in future experiments, was found to be ca. 650°C. In Figure 3, the timedependent behavior of CO, VOC (Org.-C / THC) and aromatics (sum) has been depicted. These concentrations are recorded for four burning cycles of the reference test (Table 1).

6.3.2 INTEGRATION OF UNCOATED AL_2O_3-FOAMS AS A SUPPORT MATERIAL

Al_2O_3-foams were tested as a possible support material for a suitable catalyst at the start of the experimental stage. For this purpose, two such foams were inserted into the walls of the combustion chamber. However, first of all it was important to calculate the pressure drop across the monoliths in order to ascertain the smooth operation of the stove after installing the monoliths. The pressure drop across the monoliths was found to be lower than 0.5 Pa which is sufficiently low and shows the applicability of the foams.

As observed, there is no negative effect on the combustion behavior of the stove after installing the uncoated Al_2O_3-foams, so it leads to the testing of the monoliths in the combustion chamber with Mixed Metal Oxide (MMO) as an active phase.

TABLE 1: Emission values during the reference test

Pollutants	Emission values mg/m³ i.N., 13 % O_2
CO	1514
VOC	132
Aromatics (sum)*	26
Dust	37

A total of 15 aromatic compounds, the important including benzene, naphthalene and toluene

TABLE 2: Reduction in emissions after integrating MMO/Al$_2$O$_3$ foams

Experiment Unit	Reference mg/m³ i.N., 13 % O_2	MMO/α-Al$_2$O$_3$ mg/m³ i.N., 13 % O_2	Reduction %
CO	1514	1201	21
VOC (Org.-C, FID)	109	63	42
VOC (Org.-C, FTIR)	132	83	37
dust with rinsing	37	17	55
dust without rinsing	33	14	57

TABLE 3: Emission reduction after integrating MMO/α-Al$_2$O$_3$-foams with heat reflecting plate.

Experiment Unit	Reference mg/m³ i.N., 13 % O_2	MMO/α-Al$_2$O$_3$ mg/m³ i.N., 13 % O_2	Reduction %
CO	1514	578	62
VOC (Org.-C, FID)	109	16	85
VOC (Org.-C, FTIR)	132	35	74
dust with rinsing	37	11	71
dust without rinsing	33	10	70

6.3.3 INTEGRATION OF THE CATALYST IN THE WALLS OF LOWER COMBUSTION CHAMBER

Wall catalyst based on MMO/Al$_2$O$_3$-foam: As evident from the Table 2, after the catalyst incorporation, the emissions of CO and VOC (Org.-C)

were reduced by 21% and 42% respectively (in comparison to the reference test). Moreover, the dust emissions were also abated by 55%.

Reduction of pollutants with the integration of wall catalysts and heat reflecting plate: In order to lower the emissions, the temperature of wall catalysts in the lower combustion chamber was increased by placing a heat reflecting plate (made of vermiculite) in front of the door in the lower combustion chamber (Table 3).

Integration of the MMO/α-Al$_2$O$_3$ catalyst synthesized through Technique 1: After recording positive results concerning emission control by using a suitable catalyst, the active phase of mixed metal oxide (as used in previous experiments) was brought onto the aluminium oxide foam through a novel technique, which is termed here as "Technique 1" (described in the section 2.2).

As can be seen from the Table 4, the emissions of CO and Org.-C were reduced by 58%, clearly indicating the suitability of both the active phase and the corresponding synthesis route.

Integration of the mixed metal oxide/α-Al$_2$O$_3$ catalyst synthesized through Technique 2: On experimental basis, another technique, "Technique 2" (described in the section 2.2), has been adopted to observe the suitability of the procedure regarding better oxidation activity of the catalyst.

As evident from the Table 5, the selected synthesis route was not proved to be fruitful, as emission values were higher than using "Technique 1" (Table 6).

Aging behavior of the wall catalyst MMO/α-Al$_2$O$_3$: For the determination of the thermal and chemical deactivation of the catalyst, it was aged by fitting into a downdraft stove and subjected to real operating conditions for 630 h (equal to one heating period). The longterm/ aging experiments were planned in such a way that the catalyst was exposed to real operating conditions for three weeks (except the first aging cycle was 6 weeks) and after that immediately tested for its activity. Shortly after, the catalyst was again subjected to a long-term experiment for three weeks before being analyzed again for its stability. The results have indicated that, as shown in Table 7, the catalyst showed initially quite a promising oxidation of pollutants namely, carbon monoxide, volatile organic compounds and dust (particulate matter). This behavior can be attributed to the thermal activation of the catalyst caused by the diffusion of active phase species into the support material, resulting into the synthesis of more active catalytic phase [5]. However, as clear from Table 7, the activity of the catalyst dwindled with the passage of time. This can be possibly due to the poisoning of the active phase on the support material. However, there is so far no evidence for the provided assumptions as catalyst characterization (e.g. XRD, XPS) is planned to be done at the end of the aging experiments (after the fifth cycle).

TABLE 4: Reduction in the emissions after integrating the catalyst MMO/α-Al2O3 synthesized through Technique 1.

Experiment Unit	Reference* mg/m³ i.N., 13 % O_2	MMO/α-Al$_2$O$_3$ mg/m³ i.N., 13 % O_2	Reduction %
CO	1718	725	58
VOC (Org.-C, FID)	156	65	58
VOC (Org.-C, FTIR)	202	92	54

The reference experiment was performed again with the new batch of same fuel type

TABLE 5: Emission values after fitting the catalyst (MMO/α-Al$_2$O$_3$) synthesized through Technique 2.

Experiment Unit	Reference mg/m³ i.N., 13 % O_2	MMO/α-Al$_2$O$_3$ mg/m³ i.N., 13 % O_2	Reduction %
CO	1718	1359	21
VOC (Org.-C, FID)	156	115	26
VOC (Org.-C, FTIR)	202	147	27

Aging behavior of the wall catalyst synthesized through Technique 2: In order to get verification about thermal activation in case of mixed metal oxide catalyst, another long-term/aging experiment was performed with a selected wall catalyst, as tested earlier (see section 3.3.4), where the catalyst was exposed to real conditions in the stove for about 4.5 h. As can be seen from the Table 8, there is quite a substantial amount of reduction in the emissions. The emissions of CO and VOC (Org.-C) were reduced by 62% and 77% respectively. Clearly, there is a thermal activation effect which can be observed in regard to the selected MMO/Al$_2$O$_3$ catalyst. However, like pointed out earlier, a catalyst characterization has to be done in order to support this assumption but it is very obvious that there exists quite a high probability of thermal activation, as can be observed from multiple experimental results.

TABLE 6: Comparison between the two selected synthesis routes.

Experiment Unit	Technique 1 mg/m³ i.N., 13 % O$_2$	Technique 2 mg/m³ i.N., 13 % O$_2$
CO	725	1359
VOC (Org.-C, FID)	65	115
VOC (Org.-C, FTIR)	93	147

TABLE 7: Emission values during the course of the aging experiments with MMO/α-Al$_2$O$_3$ catalyst.

Experiment Unit	Reference mg/m³ i.N., 13 % O$_2$	after cycle-1 mg/m³ i.N., 13 % O$_2$	after cycle-2 mg/m³ i.N., 13 % O$_2$	after cycle-3 mg/m³ i.N., 13 % O$_2$
CO	1718	586	222	837
VOC (Org.-C, FID)	156	36	8	64
dust (after rinsing)	37	11	9	16

TABLE 8: Reduction in the emissions after the catalytic treatment during the "normal" and "long-term" experiments.

Experiment Unit	During normal experiment mg/m³ i.N., 13 % O$_2$	During long-term experiment mg/m³ i.N., 13 % O$_2$	Reduction %
CO	1359	518	62
VOC (Org.-C, FID)	115	27	77

6.4 FUTURE WORK AND CONCLUSIONS

The selected monoliths, primarily composed of aluminum oxide (Al_2O_3) were coated with mixed metal oxide as an active catalytic phase and later inserted into the walls of the stove in the lower combustion chamber. These Al_2O_3 foams (porosity of 10 ppi) consist of 92% α-Al_2O_3 along with the trace phases of mullite and cordierite. The results revealed that the catalyst was found to be quite active in terms of oxidation of harmful pollutants e.g. CO and VOC. In addition, two different synthesis routes for mixed metal oxides on the alumina foam were discussed. It was found that, the "Technique 1" proved to be promising as the catalyst showed higher emission reductions, as compared to the one synthesized through the "Technique 2". Perhaps, it can be attributed to the comparatively high temperature handling of the precursor, thus allowing a more mature crystallization of the active phase structure. Furthermore, the aging experiments were performed with three different wall catalysts, each consisting of mixed metal oxides but synthesized via different methods. It is quite obvious that each of the three catalysts showed a "thermal activation effect" during the long-term/aging experiments, but this assumption cannot be yet supported due to the lack of catalyst characterization, which is planned to be carried out as soon as possible.

REFERENCES

1. Demirbas A (2009) Biofuels. Springer.
2. Junginger HM, Jonker JGG, Faaij A, Cocchi M, Hektor B, et al. (2011) Summary, synthesis and conclusions from IEA Bioenergy Task 40 country reports on international bioenergy trade. IEA Bioenergy.
3. Jokiniemi J, Hytönen K, Tissar J, et al. (2008) Biomass combustion in residential heating: Particulate measurements, sampling, and physicochemical and toxicological characterization. Finland: University of Kuopio, Fine Particle and Aerosol Technology Laboratory. 92 pages.
4. Kelz J, Brunner T, Obernberger I, Jalava P, Hirvonen M (2010) PM emissions from old and modern biomass combustion systems and their health effects. 18th European Biomass Conference and Exhibition 1231-1243.
5. Tsybulya SV, Kryukova GN, Kriger TA, Tsyrulnikov PG (2003) Structural Aspect of a Thermal Activation Effect in the MnOx/γ-Al2O3 System. Kinetics and Catalysis 44: 287-296.

CHAPTER 7

Catalytic Constructive Deoxygenation of Lignin-Derived phenols: New C–C Bond Formation Processes from Imidazole-Sulfonates and Ether Cleavage Reactions

STUART M. LECKIE , GAVIN J. HARKNESS,
AND MATTHEW L. CLARKE

The production of cellulose-derived chemicals is significantly more commercially attractive if economic value can be obtained from the lignin fraction of ligno-cellulose. There is currently great interest in researching the conversion of lignin to aromatics and alkanes. [1] These studies generally focus on the possible production of fuels, bulk, or commodity chemicals. The reactions used are depolymerisation of lignin, and hydro-deoxygenation reactions i.e. the replacement of the C–O bond with inert C–H bonds. [2,3]

We considered a new challenge in this field of renewable chemistry; if a small portion of lignin-derived bio-oils can be converted into one or more higher value fine chemicals, prior to hydro-deoxygenation, then extra economic value can be derived from this lignin fraction. The research to find efficient lignin depolymerisation methods is still very much an expanding effort. None-the-less in the research published so far, 2-methoxyphenol (guaiacol) is a very common major component in lignin-derived bio-oils.3 2-Methoxyphenol is somewhat more volatile than some aromatic components, and can also be converted during processing to catechol, which may be possible to separate due to its acidity. While other building blocks may become viable in the future, it seems likely that 2-methoxyphenol and catechol will be produced from lignin feedstocks. [2b,3] Another possibility is that catalysis chemistry could be developed to selectively remove guaiacol or other monomers from lignin. [4] A further speculative possibility is to functionalise specific monomers in a bio-oil mixture, to give new fine chemicals that are readily separated from the rest of the bio-oil.

In order to give a larger range of possible target fine chemicals, new catalytic chemistry needs to be developed to convert chemicals like 2-methoxyphenol into less oxygenated, but still functionalised aromatic compounds, i.e. the challenge of catalytic constructive deoxygenation (Scheme 1). Longer term requirements are likely to be heavily focused on cost, so while improving the economics of the catalytic processes needs to be addressed in due course, certain aspects such as the reagents used to activate C–O bonds and the processes chosen to study need to considered now. This actually leads to some interesting problems for catalysis chemists to study. Here we show the first studies on this concept and report new protocols to replace one or both C–O bonds in 2-methoxyphenol with C–C bonds.

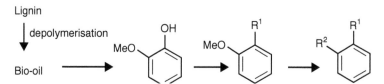

SCHEME 1: Catalytic constructive deoxygenation of 2-methoxyphenol might allow this lignin-derived feedstock to be used to produce less oxygenated aromatic compounds (R1/ R2 = various alkyl/aryl groups).

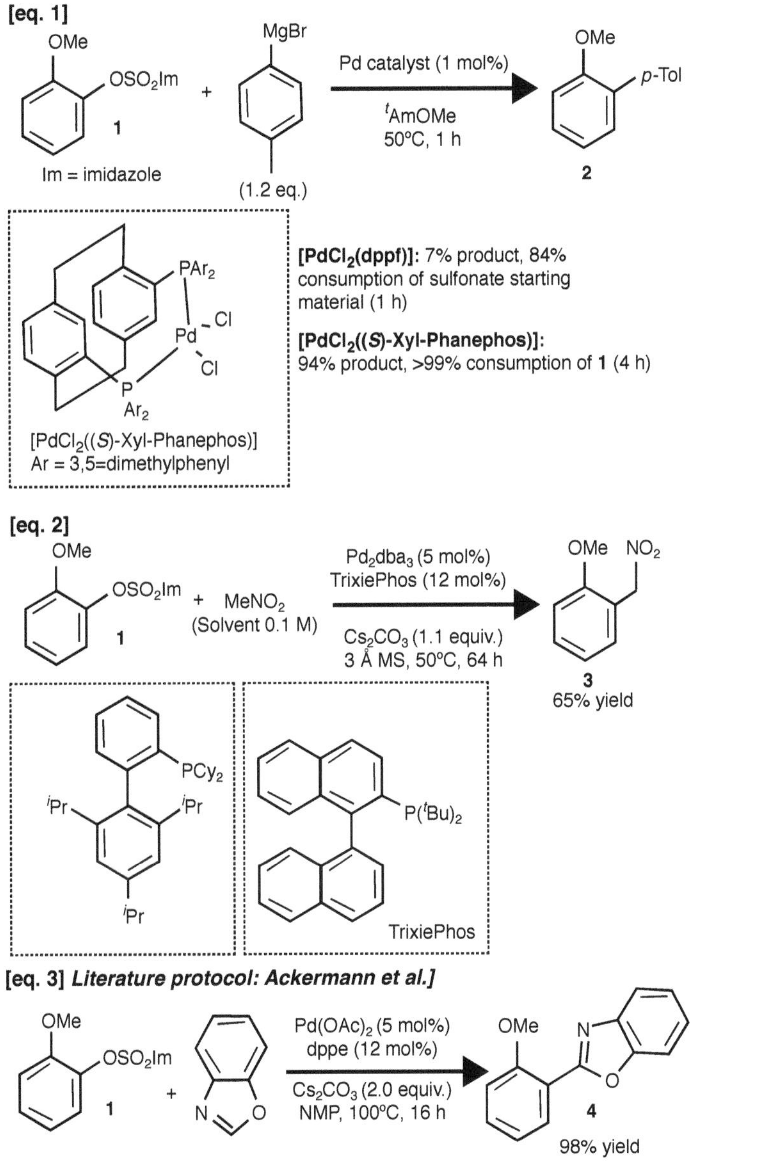
[eq. 1]
OMe
OSO2Im
1
Im = imidazole
+
MgBr
(1.2 eq.)
Pd catalyst (1 mol%)
tAmOMe
50°C, 1 h
OMe
p-Tol
2
PAr2
Pd
Cl
Cl
P
Ar2
[PdCl2((S)-Xyl-Phanephos)]
Ar = 3,5=dimethylphenyl
[PdCl2(dppf)]: 7% product, 84% consumption of sulfonate starting material (1 h)
[PdCl2((S)-Xyl-Phanephos)]: 94% product, >99% consumption of 1 (4 h)
[eq. 2]
OMe
OSO2Im
1
+ MeNO2
(Solvent 0.1 M)
Pd2dba3 (5 mol%)
TrixiePhos (12 mol%)
Cs2CO3 (1.1 equiv.)
3 Å MS, 50°C, 64 h
OMe NO2
3
65% yield
PCy2
iPr
iPr
iPr
P(tBu)2
TrixiePhos
[eq. 3] Literature protocol: Ackermann et al.]
OMe
OSO2Im
1
+
N
O
Pd(OAc)2 (5 mol%)
dppe (12 mol%)
Cs2CO3 (2.0 equiv.)
NMP, 100°C, 16 h
OMe
N
O
4
98% yield

SCHEME 2: Catalytic reactions of 2-methoxyphenyl-1H-imidazole-1-sulfonate with Grignards, nitromethane and benzoxazole.

The conversion of phenolic derivatives to activated compounds, followed by cross-coupling reactions is, of course, known methodology in a general sense. However, specific methods need to be developed for 2-methoxyphenol that do not use expensive triflates or other incompatible or expensive reagents.

We have focused on the coupling of the imidazole-sulfonate of 2-methoxyphenol, 1 (Scheme 2) since this is a reasonably cheap leaving group that is also claimed to give less toxic waste streams relative to triflates and their derivatives. [5] For the cross-coupling partners, we have assessed a range of suitable possibilities, but here we have studied Grignard reagents, nitromethane, heteroaromatic compounds and the cyanide anion since these are economic coupling partners.

The Kumada cross-coupling of Grignard reagents with imidazole-sulfonates had not been reported, but our starting point was procedures that work well for aryl halides. The use of [PdCl$_2$(dppf)] (dppf = 1,1'-bisdiphenylphosphino-ferrocene) in methyl-THF has previously been found to be an excellent procedure for Grignard cross-coupling, even under very concentrated conditions. [6] However, none of the desired product was formed. Changing solvent to tert-amyl methyl ether enabled the cross coupling to proceed, although very unselectively, and very slowly using [PdCl$_2$(dppf)] (see ESI†). We were pleased to find that the use of [PdCl$_2$((S)-Xyl-phanephos)] [7] as catalyst is much more active and selective (Scheme 2, eqn (1)).

Excellent results were obtained using this previously unexplored Grignard coupling catalyst. In the ESI† a further table of results comparing Pd/dppf and Pd/Xyl-phanephos for aryl halides shows that PdCl$_2$(Xyl-phanephos) is a more active catalyst than the widely applied Pd/dppf catalyst for this reaction. We used the expensive enantiomerically pure catalyst, but the racemic analogue would be a relatively economic ligand to use in achiral C–C bond forming reactions. Xyl-phanephos has a larger bite angle than dppf; this is generally associated with more efficient reductive elimination, but a lower propensity to other off-cycle events using this system is also possible.

The coupling of nitromethane with imidazole-sulfonates was not known, but there had been a report of coupling aryl halides with nitromethane. [8] This protocol makes use of Pd/XPhos as the leading catalyst, although none of these operates at low catalyst loading. Our initial screening (see ESI† and Scheme 2, eqn (2)) revealed that the combination used for aryl halides was ineffective, but Pd/TrixiePhos proved to be the only reasonably effective catalyst for coupling nitromethane with 1 to give desired product 3. Further research on more active nitromethylation catalysts in general would be worthwhile. It seems likely a mono-ligated Pd centre is desirable given the very bulky ligands required for any conversion in this study. Another pro-nucleophile with relatively acidic C–H bonds is benzoxazole. In this case, there had already been a report of C–H functionalisation using a range of phenols activated as their imidazole sulfonates, including compound 1. [9] We therefore used this procedure here (Scheme 2, eqn (3)), although again note that improvements in catalytic turnover are desirable in the future.

We next studied cyanation using cheap and non-toxic $K_4Fe(CN)_6$. This cyanide source, introduced in a seminal paper by Beller and co-workers, has been studied in the coupling with aryl halides and aryl mesylates, but not imidazole-sulfonates. [10] A more extensive screening of conditions and many different catalysts can be found in the ESI;† most catalysts are not sufficiently active, and 1 slowly hydrolyses to guaiacol, 6. Table 1 shows that combinations of either X-Phos or triphenylphosphine combined with Pd(II) pre-catalysts were effective. The more economic triphenylphosphine based system was found to give the optimal results for the production of 5 (Table 1, entry 7).

The cross-coupling processes shown above suggest that, beyond the realm of this specific project, it should be possible to carry out effective cyanation, Grignard cross-coupling and nitromethylation reactions using phenol-imidazolesulfonates and the new procedures identified here. Moreover, these studies show that, with further research, it should be feasible to develop scalable methods for the C–C bond forming reaction using 1. This should be useful for making various phenolic compounds containing only one aromatic C–O bond. To increase the potential scope of this building block, it would be desirable to be able to swap the remaining aromatic C–O bond for a C–C bond. [11] There are some important fine chemicals that could be produced effectively using this type of route, [12] but at this early stage, we wanted to map out what was possible.

TABLE 1: Selected examples from the optimisation of the cyanation of 2-methoxyphenyl-1H-imidazole-1-sulfonate

Entry	Pd precursor (mol%)	Ligand (mol%)	Temp. (°C)	Time (h)	Ration 5:1:6[a]	Yield (%)
1[b]	Pd(OAc)$_2$ (5 mol%)	X-Phos (10 mol%)	110	72	99:0:1	50
2[c]	Pd(TFA)$_2$ (1 mol%)	X-Phos (2 mol%)	110	48	68:25:7	—
3	Pd(TFA)$_2$ (1 mol%)	X-Phos (2 mol%)	110	48	81:8:11	—
4	Pd(TFA)$_2$ (1 mol%)	X-Phos (3 mol%)	110	24	88:0:12	—
5[c]	Pd(TFA)$_2$ (1 mol%)	PPh3 (3 mol%)	110	24	88:8:4	—
6	Pd(TFA)$_2$ (1 mol%)	PPh3 (3 mol%)	100	48	95:0:5	—
7[c]	Pd(TFA)$_2$ (1 mol%)	PPh3 (3 mol%)	90	48	94:5:1	73

[a]As judged by 1H NMR of the crude reaction mixture. Yields are pure product after chromatography. [b]0.5 equiv. of K4FeCN6. [c]0.21 equiv. of K_4FeCN_6.

As already noted, it is convenient to produce 4, using C–H activation coupling of benzoxazole with 1, so we considered modifying the Meyers reaction towards this class of substrate. The Meyers reaction normally uses certain oxazolines as activating groups for ether cleavage, [13] and to the best of our knowledge, there are not any examples of Meyers coupling using this type of benzoxazole. We were pleased to find that these reactions proceed well at near ambient temperatures using a range of aromatic, alkenyl and alkyl Grignards. Scheme 3 lists the products 7a–7h produced and reaction conditions.

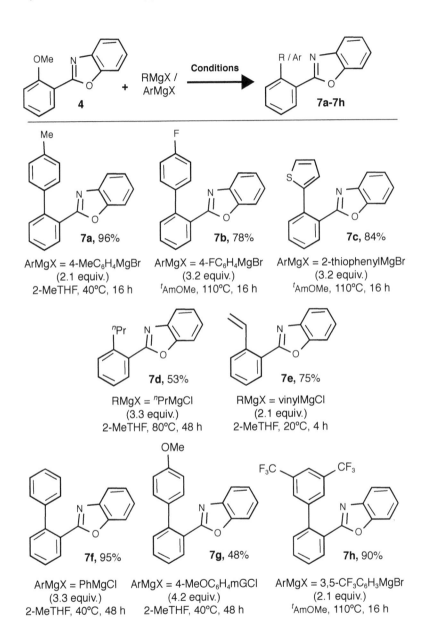

SCHEME 3: Modified Meyers reaction of anisyl-benzoxazoles with Grignard reagents.

In summary, some cross-coupling reactions that use relatively economic nucleophilic partners and the imidazole-sulfonate of 2-methoxyphenol, 1 have been studied. It is proposed that this type of catalysis might be useful for creaming off some high value products from bio-oil mixtures, or bio-oil derived 2-methoxyphenol. In this case, we have identified several new protocols for cross-coupling imidazole sulfonate derivatives with Grignards, nitromethane and a non-toxic cyanide source. Modified Meyers reactions on benzoxazoles are also reported. These discoveries should prove enabling to those needing new organic methodology, in addition to presenting the first steps towards constructive deoxygenation reactions of renewables.

We thank the EPSRC for funding, and all the technical staff in the School of Chemistry for their assistance.

REFERENCES

1. J. Zakzeski, P. C. A. Bruijnincx, A. L. Jongerius and B. M. Weckhuysen, Chem. Rev., 2010, 110, 3552
2. (a) R. C. Runnebaum, T. Nimmanwudipong, D. E. Block and B. C. Gates, Catal. Sci. Technol., 2012, 2, 113 RSC; (b) M. P. Pandey and C. S. Kim, Chem. Eng. Technol., 2011, 34, 29
3. (a) C. Amen-Chen, H. Pakdel and C. Roy, Biomass Bioenergy, 1997, 13, 25 Cross-Ref CAS Search PubMed; (b) J. Zakzeski, A. L. Jongerius, P. C. A. Bruijnincx and B. M. Weckhuysen, ChemSusChem, 2012, 5, 1602; (c) P. Varanasi, P. Singh, M. Auer, P. D. Adams, B. A. Simmons and S. Singh, Biotechnol. Biofuels, 2013, 6, 14; (d) For oxidative cleavage, see R. Jastrzebski, B. M. Weckhuysen and P. C. A. Bruijnincx, Chem. Commun., 2013, 49, 6912 RSC.
4. (a) A. G. Sergeev, J. D. Webb and J. F. Hartwig, J. Am. Chem. Soc., 2012, 134, 20226; (b) T. vom Stein, T. Wiegard, C. Merkens, J. Klankermayer and W. Leitner, ChemCatChem, 2013, 5, 439
5. (a) L. Ackermann, R. Sandmann and W. Song, Org. Lett., 2011, 13, 1784; (b) J. F. Cívicos, D. A. Alonso and C. Nájera, Eur. J. Org. Chem., 2012, 3670; (c) J. Albaneze-Walker, R. Raju, J. A. Vance, A. J. Goodman, M. R. Reeder, J. Liao, M. T. Maust, P. A. Irish, P. Espino and D. R. Andrews, Org. Lett., 2009, 11, 1463
6. (a) T. Hayashi, M. Konishi, Y. Kobori, M. Kumada, T. Higuchi and K. Hirotsu, J. Am. Chem. Soc., 1984, 106, 158; (b) T. Hayashi, M. Konishi and M. Kumada, Tetrahedron Lett., 1979, 21, 1871; (c) E. J. Milton and M. L. Clarke, Green Chem., 2010, 12, 381 RSC.

7. T. M. Konrad, J. A. Fuentes, A. M. Z. Slawin and M. L. Clarke, Angew. Chem., Int. Ed., 2010, 49, 9197

8. (a) R. R. Walvoord, S. Berritt and M. C. Kozlowski, Org. Lett., 2012, 14, 4086; (b) R. R. Walvoord and M. C. Kozlowski, J. Org. Chem., 2013, 78, 8859

9. L. Ackermann, S. Barfusser and J. Pospech, Org. Lett., 2010, 12, 724

10. (a) T. Schareina, A. Zapf and M. Beller, Chem. Commun., 2004, 1388 RSC; (b) P. Anbarasan, T. Schareina and M. Beller, Chem. Soc. Rev., 2011, 40, 5049 RSC and references cited therein; (c) P. Y. Yeung, C. M. So, C. P. Lau and F. Y. Kwong, Angew. Chem., Int. Ed., 2010, 49, 8918

11. (a) J. W. Dankwardt, Angew. Chem., Int. Ed., 2004, 43, 2428; (b) M. Tobisu, T. Shimasaki and N. Chatani, Angew. Chem., Int. Ed., 2008, 47, 4866

12. A particularly desired problem to solve would be the coupling of some form of para-tolyl nucleophile with guaiacol-derived 2-methoxybenzonitrile. Such a reaction gives 4'-methyl-[1,1'-biphenyl]-2-carbonitrile, a key building block for the production of Losartan. Such compounds of relatively high value and significant demand are particularly important targets for this type of approach.

13. (a) J. Mortier, Curr. Org. Chem., 2011, 15, 2413; (b) M. Reuman and A. I. Meyers, Tetrahedron, 1985, 41, 837; (c) T. G. Gant and A. I. Meyers, Tetrahedron, 1994, 50, 2297.

CHAPTER 8

Efficient One-Pot Synthesis of 5-Chloromethylfurfural (CMF) from Carbohydrates in Mild Biphasic Systems

WENHUA GAO, YIQUN LI, ZHOUYANG XIANG, KEFU CHEN, RENDANG YANG, AND DIMITRIS S. ARGYROPOULOS

8.1 INTRODUCTION

Locating new and versatile platform chemicals and biofuels from sustainable resources to replace those derived from petrochemicals is a central ongoing and urgent task prompted by depleting fossil fuel reserves and growing global warming concerns [1–4]. Alternative fine chemicals and biofuels that have been suggested to address some of these issues are butanol [5], ethanol [5], dimethylfuran [1], 5-ethoxymethylfurfural [2], γ-valerolactone, and alkanes produced from biomass [6,7]. Many of these alternatives rely on the efficient conversion of biomass carbohydrates into furfural derivatives. This is because biomass carbohydrates constitute 75% of the World's renewable biomass and cellulose [4] and as such, they represent a promising alternative energy and sustainable chemical feed-

Efficient One-Pot Synthesis of 5-Chloromethylfurfural (CMF) from Carbohydrates in Mild Biphasic Systems. © Gao W, Li Y, Xiang Z, Chen K, Yang R, and Argyropoulos DS. Moleculres **18** *(2013), doi:10.3390/molecules18077675. Licensed under Creative Commons Attribution 3.0 Unported License, http://creativecommons.org/licenses/by/3.0/.*

stock. In this regard, 5-halomethyfurfurals such as 5-chloromethylfurfural (CMF) and 5-bromomethylfurfural (BMF) has received significant attention as platform chemicals for synthesizing a broad range of chemicals and liquid transportation fuels [8,9].

CMF and BMF are extremely reactive [9] so that when subjected to further chemistries the provide a variety of important compounds for fine chemicals, pharmaceuticals, furan-based polymers and biofuels. These compounds include hydroxymethylfurfural (HMF) [9,10], 2,5-dimethylfuran (DMF) [1], and 5-ethoxymethylfurfural (EMF) [2], and some biologically active compounds [11]. Among them, DMF and EMF stand out since they possess excellent properties, including high energy density, high boiling point and water stability. For these reasons, they have been promoted as novel biofuels. In particular, EMF has been the subject of considerable attention since it possesses an energy density of 8.7 kWhL^{-1}, substantially higher than that of ethanol (6.1 kWhL^{-1}), and comparable to that of standard gasoline (8.8 kWhL^{-1}) and diesel fuel (9.7 kWhL^{-1}) [12]. Although, CMF and BMF themselves are not biofuels, they could readily be converted into EMF biofuels in ethanol in nearly quantitative yields.

The conventional synthesis of CMF involves the treatment of HMF or cellulose with dry hydrogen halide. More specifically, the hydroxyl group in HMF undergoes a facile halogen substitution reaction. Examples in the literature include those of Sanda et al. who obtained CMF from the reaction of ethereal gaseous hydrogen chloride with HMF [13]. Furthermore, while the conversion of cellulose into CMF was low (12%) [14,15], a substantially higher yield (48%) was obtained for the preparation of BMF when dry HBr was employed [16]. Considering the importance of these compounds, Mascal et al. recently reported the synthesis of CMF from cellulose treated by HCl-LiCl and successive continuous extraction [2]. Unfortunately, 5-(chloromethyl)furfural, 2-(2-hydroxyacetyl)furan, 5-(hydroxylmethyl), furfural and levulinic acid were also produced with this system. More recently, Kumari et al. reported the preparation of BMF from cellulose by a modified procedure using HBr-LiBr involving continuous extraction [17]. Despite the numerous efforts aimed at these transformations, each of them suffers from at least one of the following limitations: diverse by-products in significant yields that reduce the selectivity of the reaction and its economics, low conversions and yields, harsh reaction conditions (dry hydrogen halide, relative high temperature), requirements for large amounts of costly reagents (LiCl, LiBr), prolonged reaction times and tedious operations with complex set ups (continuous extraction) [18]. These drawbacks seriously hamper their potential industrial applications. Consequently, as part of our program aimed at developing new biofuels and fine chemicals based on biomass, we embarked our research for the development of efficient and economical methods aimed at converting carbohydrates to CMF under mild reaction conditions.

In this communication we demonstrate the use of the biphasic mixture HCl-H$_3$PO$_4$/CHCl$_3$ for the one-pot conversion of carbohydrates into CMF. The rational for the use of this biphasic approach is based on the thinking that as CMF is generated from HMF it is immediately transported and extracted from the aqueous acidic phase into the organic phase significantly minimizing by-product yields [9].

TABLE 1. The effect various reaction variables on CMF yields from D-fructose[a].

Entry	HCl/H$_3$PO$_4$ (v/v)	Temperature (°C)	Time (h)	Yields (mol%)[b]
1	1/0	45	20	28.4
2	2/1	45	20	36.9
3	3/1	45	20	42.1
4	4/1	45	20	46.8
5	5/1	45	20	45.5

[a]D-fructose (5.0 mmol) was added in a mixture with specific volume ratio of 37% HCl and 85% H$_3$PO$_4$ (5.0 mL), and CHCl3 (5.0 mL). The system was stirred continuously at 45 °C for 20 h. [b]Isolated yields based on D-fructose.

8.2 RESULTS AND DISCUSSION

Initial efforts using the biphasic treatment (HCl(37%)-H$_3$PO$_4$(85%)/CHCl3) of the monosaccharide D-fructose at 45 °C for 20 h, offered a CMF yield of 35.8%. we then attempted to further optimize this result by systematically varying the volume ratio of HCl and H$_3$PO$_4$ (Table 1), the nature of the non-aqueous solvent, the reaction time and temperature. This is due to the fact that Table 1 deals only with the volume ratio of HCl and H_3PO_4. The effects of changing other variables are reported in Tables elsewhere.

TABLE 2: Effect of reaction temperature on the yields of CMF at different ratios of HCl/H_3PO_4[a].

Entry	HCl/H_3PO_4 (v/v)	Temperature (°C)	Time (h)	CMF Yield (mol %)[b]
1	3/1	35	20	38.5
2	3/1	45	20	42.1
3	3/1	55[c]	20	38.8
4	4/1	35	20	43.9
5	4/1	45	20	46.8
6	4/1	55	20	42.8
7	5/1	35	20	43.2
8	5/1	45	20	45.5
9	5/1	55	20	40.2

[a]D-fructose (5.0 mmol) was added in a mixture with specific volume ratio of HCl(37%) and H_3PO_4 (85%) (5.0 mL), and $CHCl_3$ (5.0 mL). The system stirred continuously at the special temperature for 20h. [b]Isolated yields based on D-fructose. [c]Slightly below the $CHCl_3$-water azeotrope boiling point (56.1 °C).

The data of Table 1 indicates that the amount of hydrochloric acid in the biphasic medium plays an important role in this transformation. Increasing the amount of HCl in the system seemed to concomitantly increase the yield of CMF (Table 1, entries 1–5). Even at lower volume ratios of HCl/H_3PO_4 such as 2:1, the yield of CMF was seen to increase by 8.5%, compared with the reaction with just hydrochloric acid (Table 1, entries 1,2). H_3PO_4 offered enough hydrogen ions to promote the fructose selective conversion to CMF. The D-fructose molecule needs catalytic hydrogen ions to form fructofuranosyl intermediates, and the acid-induced elimination of three moles of water from this intermediate leads to the conversion to 5-hydroxymethyl-2-furaldehyde (HMF) [19,20]. As hydrochloric acid existed in the system, HMF could easily be transformed to 5-chloromethylfurfural (CMF). Increasing the mixed acid ratio (HCl/H$_3$PO$_4$) to 3:1 offered a CMF yield of 42.1% (Table 1, entry 3). Further increases in the mixed acid ratio (HCl/H$_3$PO$_4$) offered only marginal yield increases in the CMF yield (Table 1, entries 4,5). Therefore, in the dehydration reaction to produce CMF from fructose, HCl performed as a good mineral acid catalyst and provided the essential chloride for the production of CMF. Román-Leshkov et al. also observed various acid catalysts used to implement the dehydration reaction, while HCl showed the highest catalytic ability amongst common mineral acids [21].

The effect of reaction temperature was examined in detail by conducting CMF yield studies at 35, 45, and 55 °C at different ratios of HCl/H$_3$PO$_4$ (Table 2).

Upon increasing the temperature from 35 to 45 °C, the yields of CMF were seen to increase, when the mixed acid ratio of HCl/H$_3$PO$_4$ was varied from 3:1, 4:1 and 5:1 (Table 2, entries 1–9). At the temperature of 55 °C, however, all the CMF yields decreased somewhat compared to 45 °C (Table 2, entries 1–9). This could be rationalized on the basis of different energies of activation for the different side-reactions being more pronounced at the slightly higher temperatures.

In the presence of acid catalysts, fructose could dehydrate to produce CMF and various by-products. Increasing temperatures improved the yields of by-products. The cyclic fructofuransyl intermediate pathway degraded by means of secondary reactions, such as fragmentation, condensation or other dehydration reactions. In this respect our data was consistent with similar research by Moreau et al. [22] and Antal et al. [23] that dealt with the D-fructose dehydration mechanism and the formation of HMF. Overall, at lower reaction temperatures, the mixed acids did not offer their catalytic role, and the yield of CMF was not high.

CMF yields obtained from the acid ratio (HCl/H$_3$PO$_4$) of 4:1 at different temperatures were seen to be higher than when the acid ratio was 3:1 or 5:1 (Table 2, entries 1–9). As such the acid volume ratio and the reaction temperature were established to be 4:1 and 45 °C, respectively, for this transformation.

The immiscible organic solvent plays an important role in extracting the products formed during the acidic aqueous phase reaction, thud significantly reducing side reactions. The nature of the extraction solvent was thus further explored and it was verified that indeed chloroform (Table 3, entry 1) was the extraction solvent of choice amongst all screened solvents. For example, 1, 2-dichlororethane (Table 3, entry 2) and 1-chlorobutane (Table 3, entry 3) were not able to extract all the CMF from the acidic aqueous phase, while toluene promoted a multitude of side reactions manifesting themselves in the complexity of the isolated product mixture. Unlike toluene, chloroform showed an excellent selectivity for extracting CMF from the acidic aqueous phase, simultaneously promoting reaction uniformity and reducing by-product diversity. Overall, this makes the work-up with chloroform and purification steps convenient with potential practical applications.

TABLE 3: Effect of extraction solvent on reaction yields[a].

Entry	Extraction Solvent	Temperature (°C)	Time (h)	CMF Yield (mol%)[b]
1	Chloroform	45	20	46.8
2	1,2-Dichloroethane	45	20	28.8
3	1-Chlorobutane	45	20	22.6
4	Toluene	45	20	--- [c]

[a]D-fructose (5.0 mmol) was added in a mixture with 4/1 (v/v) of 37% HCl (4.0 mL) and 85% H_3PO_4 (1.0 mL), and organic solvent (5.0 mL) and stirred continuously at 45 °C for 20 h. [b]Isolated yields based on D-fructose. [c]Complex mixture of compounds was formed as indicated by GC-MS.

Finally the effect of reaction time was investigated using the optimized reaction conditions determined so far. A 10 h reaction time offered a CMF yield of only 34.6% (Table 4, entry 1). Extending the reaction time to 20 h, dramatically increased the CMF yield by 12.2% (Table 4, entries 1,2). Further extension of the reaction time did not offer additional CMF yield improvements (Table 4, entries 3,4).

TABLE 4: Effect of reaction time on reaction yields a.

Entry	HCl/H$_3$PO$_4$ (v/v)	Extractant	Temperature (°C)	Time (h)	CMF Yield (mol%)[b]
1	4:1	CHCl$_3$	45	10	34.6
2	4:1	CHCl$_3$	45	20	46.8
3	4:1	CHCl$_3$	45	30	44.9
4	4:1	CHCl$_3$	45	40	45.2

[a]D-fructose (5.0 mmol) was added in a mixture with 4/1 (v/v) of 37% HCl (4.0mL) and 85% H$_3$PO$_4$ (1.0 mL), and CHCl$_3$ (5.0 mL) and stirred continuously at 45 °C for special time. [b]Isolated yields based on D-fructose.

From the above analysis, a possible route of the fructose dehydration to form CMF as illustrated in Scheme 1 can be suggested.

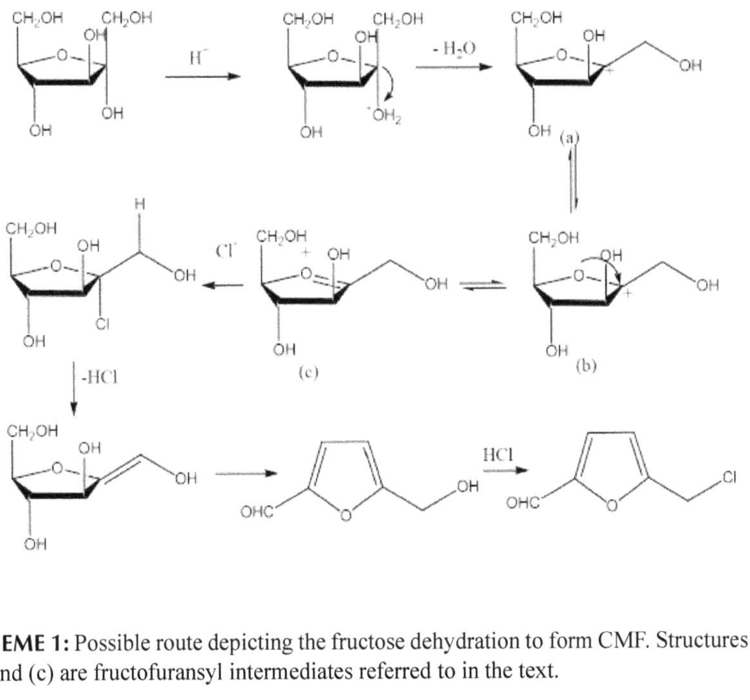

SCHEME 1: Possible route depicting the fructose dehydration to form CMF. Structures (a), (b) and (c) are fructofuransyl intermediates referred to in the text.

In an effort to further understand and chart the selectivity and general applicability of the proposed methodology a series of mono-saccharine, two disaccharides and a poly-saccharide were examined. The data of Table 5 showed that fructose offered the highest yields of CMF (Table 5, entry 1), while glucose and cellulose the lowest (Table 5, entries 2–6). Interestingly D-fructose and D-glucose, while they are chemically very similar (Scheme 2), they produced significantly different yields of CMF (Table 5, entries 1,2). This is most likely because with the keto-hexose (D-fructose) was easier to form the fructofuransyl intermediate than with the aldohexose (D-glucose), due to the higher reactivity of ketohexoses. It is likely that D-glucose could not dehydrate directly and transform into fructofuransyl intermediates under our conditions. The glucose to fructose isomerization reaction seems to be a pre-requisite for the formed fructose to dehydrate and yield CMF under the mixed acid catalytic system we propose. Our contention is supported by Rosatella et al. who proposed two methods to form the fructofuransyl intermediates, and the main one was the glucose to fructose isomerization reaction [24]. Furthermore, Zhao et al. [25] and Huang et al. [26] observed that in order to obtain high HMF yields from glucose the selective in situ isomerization of fructose was essential.

TABLE 5: Effect of different carbohydrate substrate on CMF yields[a].

Entry	Carbohydrate	CMF Yield (mol %)[b]
1	Fructose	46.8
2	Glucose	7.3
3	Sorbose	16.4
4	Sucrose	43.1
5	Cellobiose	19.3
6	Cellulose	7.8

[a] *all reactions were performed in HCl-H3PO4/CHCl3 biphasic system at 45 °C for 20 h.* [b] *isolated yield based on carbohydrate.*

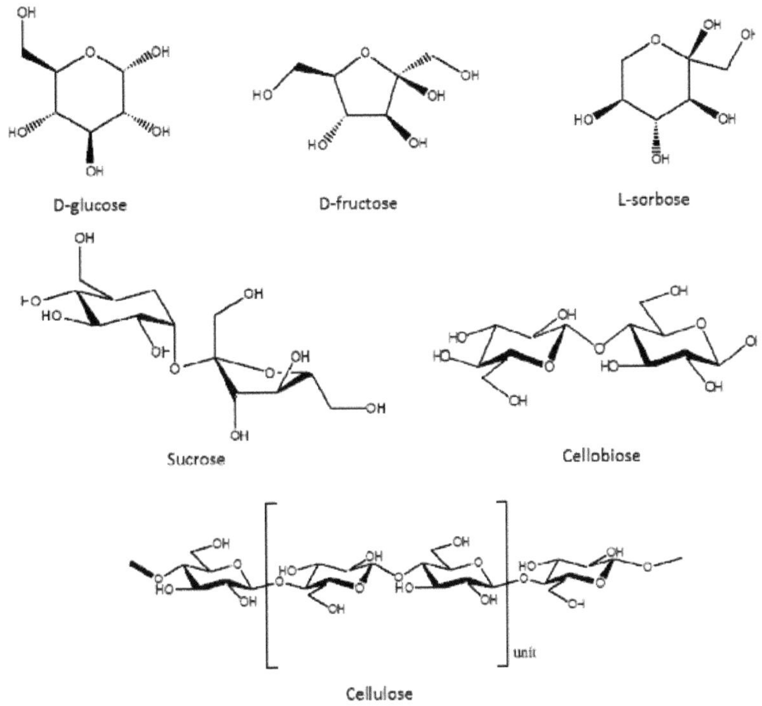

D-glucose   D-fructose   L-sorbose

Sucrose   Cellobiose

Cellulose

SCHEME 2: Structures of carbohydrates used and discussed in this work.

The monosaccharide sorbose offered a better CMF yield than D-glucose (Table 5, entry 3), since it is a ketohexose. Although sorbose is one of the epimers (C-2 and C-3) of D-fructose (Scheme 2), it had similar reactivity to D-glucose. Our data is further supported by that of Khajavi et al. who has also documented that sorbose and glucose showed almost the same ability to produce HMF, but much lower than fructose [27].

The disaccharide sucrose (Scheme 2) comprised of linked fructose and glucose units afforded CMF yields similar to fructose (Table 5, entry 4, 43.1 mol%). In the mixed acid aqueous medium, the sucrose molecule was quickly hydrolyzed into fructose and glucose. Almost all of the fructose could be selectively converted into CMF in the biphasic system, due to the higher CMF yield from sucrose. Literature data also reports that 90% fructose solutions (obtained from sucrose hydrolysis) can be converted into fructofuransyl intermediates, while the glucose residue remains nearly unchanged [28]. Therefore, the overall activating nature of the carbonyl group in the C-2 position of the sugar seems to be apparent. Its absence (glucose and cellulose) dramatically reduces the reaction yields toward the production of CMF.

Cellobiose consists of two glucose molecules joined by equatorial C1-C4 glycosidic bonds (Scheme 2). In an aqueous mixed acid solution, cellobiose was easily hydrolyzed to glucose molecules and as such it showed a somewhat higher CMF yield of 19.3% (Table 5, entry 5), while the polysaccharide cellulose only showed similar yields as glucose (Table 5, entry 6). The possible explanation for this observation was that cellulose is a high molecular weight linear polysaccharide, and its degradation involves the breaking of bonds between glucose units within the chain. As such the mixed acidic aqueous system might produce many random chain scissions, leading to many secondary reactions. Similar data have also been addressed in the literature by Emsley et al. [29] and Scheiding et al. [30].

Finally we examined various pulps and a wood powder for their potential to prepare CMF with the proposed methodology. Hardwood kraft pulp with the highest total sugar content (Table 6), containing mostly cellulose and hemicellulose, (which can hydrolyze into glucose and xylose) offered a yield of CMF of 21.3 mol% (based on the glucose of the pulp). This was a much higher than either glucose or cellulose could offer by themselves (Table 7, entry 1; Table 5, entries 2,6). The softwood (Norway spruce) thermomechanical pulp offered CMF yields of 22.8% while the hardwood powdered wood gave a CMF yield of 31.4% (both figures based on total sugar contents of these samples) (Table 7). These yields may be explained on the basis that mannose and other hexoses present in wood may also convert to fructofuransyl intermediates besides glucose. This data is supported by earlier efforts, where it was shown that mannose and glucose was of similar reactivity transforming to fructofuransyl intermediates producing hydroxymethylfurfural (HMF) [27]. Overall, our data with the complete lignocellulosic substrate is encouraging since all wood components were present during the reaction (cellulose, lignin and hemicelluloses). Under these circumstances, we believe that the obtained yields of CMF were encouraging. Efforts in our laboratory are continuing to further promote these yields in the presence of selective catalysts in the system [31].

TABLE 6: Sugar contents present in the examined lignocellulosic materials.

Lignocellulose Sample	Rhamnose (%)	Arabinose (%)	Galactose (%)	Glucose (%)	Xylose (%)	Mannose (%)	Total Sugars (%)
Eucalyptus Kraft pulp	0.0	0.0	0.0	75.9	20.8	0.0	85.7
Norway Spruce Softwood TMP	0.0	0.0	1.8	44.2	6.3	11.8	64.1
Eucalyptus Hardwood	0.0	0.0	1.3	45.0	17.5	0.9	64.7

TABLE 7: Preparation of CMF with wood pulp and wood powder[a].

Entry	Lignocellulose Sample	CMF Yield (mol %)[b]	CMF Yield (mol %)[c]
1	Eucalyptus Kraft pulp	21.3	16.0
2	Norway Spruce softwood TMP	33.7	22.8
3	Eucalyptus hardwood	47.4	31.4

[a]Lignocellulose sample (1.0 mg) was added in a mixture with 4/1 (v/v) of 37% HCl (4.0 mL) and 85% H_3PO_4 (1.0 mL), and $CHCl_3$ (5.0 mL) and stirred continuously at 45 °C for special time at 45 °C for 20 h. [b]Isolated yield based on the glucose in the lignocellulose sample. [c]Isolated yield based on the total sugars in the lignocellulose sample.

8.3 EXPERIMENTAL

8.3.1 MATERIALS AND INSTRUMENTS

All solvents and chemicals used were as obtained from commercial suppliers, unless otherwise indicated. 1H-NMR spectra were recorded on a

Bruker Avance 300 instrument using CDCl3 as solvent and TMS as the internal standard. GC-MS spectra were performed on an HP G1800B GCD system.

Eucalyptus globulus wood powder and its ensuing kraft pulp were examined as well as a Norway Spruce sample of unbleached thermomechanical pulp, which was sampled in a Swedish mill of approximate 38% consistency and 85 mL Canadian Standard Freeness prepared by one-stage refining and a subsequent reject refining (about 20%) stage. All wood materials used in this work represent standard samples, being the subject of Cost action E 41 entitled; "Analytical tools with applications for wood and pulping chemistry" operated by the European Union. The sugar profiling for these materials was examined according to the procedure of Min et al. followed by ion chromatography (Dionex IC-3000; Dionex, Sunnyvale, CA, USA) (Table 6) [32,33].

8.3.2 GENERAL PROCEDURE FOR THE SYNTHESIS OF CMF FROM MODEL CARBOHYDRATES

The selected carbohydrate (5.0 mmol) was added in a mixture of 37% HCl (4.0 mL), 85% H_3PO_4 (1.0 mL), and $CHCl_3$ (5.0 mL) and it was stirred continuously at 45 °C for 20 h. Then an equal volume of water (5.0 mL) was added to quench the reaction. The reaction mixture was then extracted with $CHCl_3$ 3 times. The combined organic extracts were then dried with anhydrous Na_2SO_4 for 4 h. Finally, the organic extracts were subjected to liquid chromatography (silica gel, CH_2Cl_2 as eluent) to offer the desired 5-chloromethylfurfural. The procedure of treating lignocellulose sample (Table 6) was almost the same as the carbohydrate, except adding the selected simple 1.0 mg each trial. The structure of 5-chloromethylfurfural was confirmed using 1H-NMR and GC-MS as follows: 1H-NMR ($CDCl_3$): δ = 4.60 (s, 2H), 6.58 (d, J = 3.6 Hz, 1H), 7.18 (d, J = 3.6 Hz, 1H), 9.64 (s, 1H) ppm. GC-MS (EI, 80 eV): m/z 146 (M^+, ^{37}Cl, 10.53), 144 (M^+, ^{35}Cl, 32.0), 109 ($C_6H_5O_2^+$, 100), 81 ($C_5H_5O^+$, 17.3).

8.4 CONCLUSIONS

In summary, this note describes an optimized biphasic system (HCl-H_3PO_4/CHCl$_3$) that may pave the way for the development of a simple, mild, and cost-effective protocol for the conversion of various carbohydrates to CMF. The systematic optimization effort undertaken here delineates the structural features of carbohydrate residues that offer optimum CMF yields. Overall, the described procedure offers several advantages over other methodologies including mild reaction conditions; satisfactory product yields; and a simple experimental and isolation process.

REFERENCES

1. Román-Leshkov, Y.; Barrett, C.J.; Liu, Z.; Dumesic, J.A. Production of dimethylfuran for liquid fuels from biomass-derived carbohydrates. Nature 2007, 447, 982–985.
2. Mascal, M.; Nikitin, E.B. Direct, High-yield conversion of cellulose into biofuel. Angew. Chem. Int. Ed. 2008, 47, 7924–7926.
3. Metzger, J.O. Production of liquid hydrocarbons from biomass. Angew. Chem. Int. Ed. 2006, 45, 685–698.
4. Corma, A.; Iborra, S.; Velty, A. Chemical routes for the transformation of biomass into chemicals. Chem. Rev. 2007, 107, 2411–2502.
5. Savage, D.F.; Way, J.; Silver, P.A. Defossiling fuel: how synthetic biology can transform biofuel production. ACS Chem. Biol. 2008, 3, 13–16.
6. Huber, G.W.; Chheda, J.N.; Barrett, C.J.; Dumesic, J.A. Production of liquid alkanes by aqueousphase processing of biomass-derived carbohydrates. Science 2005, 308, 446–450.
7. Bond, J.Q.; Alonso, D.M.; Wang, D.; West, R.M.; Dumesic, J.A. Integrated catalytic conversion of γ-valerolactone to liquid alkenes for transportation fuels. Science 2010, 327, 1110–1114.
8. Sanda, K.; Rigal, L.; Gaset, A. Optimisation of the synthesis of 5-chloromethyl-2-furancarboxaldehyde from D-fructose dehydration and in-situ chlorination of 5-hydroxymethyl-2-furancarboxaldehyde. J. Chem. Technolo. Biotechnolo. 1992, 55, 139–145.
9. Lewkowski, J. Synthesis, chemistry and applications of 5-hydroxymethylfurfural and its derivatives. ARKIVOC 2001, 1, 17–54.
10. James, O.O.; Maity, S.; Usman, L.A.; Ajanaku, K.O.; Ajani, O.O.; Siyanbola, T.O.; Sahu, S.; Chaubey, R. Towards the conversion of carbohydrate biomass feedstocks to biofuels via hydroxylmethylfurfural. Energy Environ. Sci. 2010, 3, 1833–1850.

11. Villain-Guillot, P.; Gualtieri, M.; Bastide, L.; Roquet, F.; Martinez, J.; Amblard, M.; Pugniere, M.; Leonetti, J.P. Structure-activity relationships of phenyl-furanyl-rhodanines as inhibitors of RNA polymerase with antibacterial activity on biofilms. J. Med. Chem. 2007, 50, 4195–4204.

12. Gruter, G.J.M.; Dautzenberg, F. Method for the synthesis of 5-alkoxymethylfurfural ethers and their use. Eur. Pat. Appli. 2007, 1834950 A1.

13. Sanda, K.; Rigal, L.; Gaset, A. Synthesis of 5-bromomethyl-2-furancarboxaldehyde and 5-chloromethyl-2-furancarboxaldehyde. Carbohydr. Res. 1989, 187, 15–23.

14. Sanda, K.; Rigal, L.; Delmas, M.; Gaset, A. The Vilsmeier reaction: A new synthetic method for 5-(chloromethyl)-2-furaldehyde. Synthesis 1992, 6, 541–542.

15. Canas, S.; Belchior, A.P.; Spranger, M.I.; Brun-de-Sousa, R. High-performance liquid chromatography method for analysis of phenolic acids, phenolic aldehydes, and furanic derivatives in brandies. Development and validation. J. Sep. Sci. 2003, 26, 496–502.

16. Hibbert, H.; Hill, H.S. Studies on cellulose chemistry II. The action of dry hydrogen bromide on carbohydrates and polysaccharides. J. Am. Chem. Soc. 1923, 45, 176–182.

17. Kumari, N.; Olesen, J.K.; Pedersen, C.M.; Bols, M. Synthesis of 5-Bromomethylfurfural from cellulose as a potential intermediate for biofuel. Eur. J. Org. Chem. 2011, 7, 1266–1270.

18. Brasholz, M.; Von Kaenel, K.; Hornung, C.H.; Saubern, S.; Tsanaktsidis, J. Highly efficient dehydration of carbohydrates to 5-(chloromethyl) furfural (CMF), 5-(hydroxymethyl) furfural (HMF) and levulinic acid by biphasic continuous flow processing. Green Chem., 2011, 13, 1114–1117.

19. Min, D.; Li, Q.; Jameel, H.; Chiang, V.; Chang, H.M. Comparison of pretreatment protocols for cellulose-mediated saccharification of wood derived from transgenic low-xylan lines of cottonwood (P. trichocarpa). Biomass Bioenergy. 2011, 35, 3514–3521.

20. Min, D.; Li, Q.; Jameel, H.; Chiang, V.; Chang, H.M. The cellulose-mediated saccharification on wood derived from transgenic low-lignin lines of black cottonwood (Populus trichocarpa). Appl. Biochem. Biotechnol. 2012, 168, 947–955.

21. Carlini, C.; Patrono, P.; Galletti, A.M.R.; Sbrana, G. Heterogeneous catalysts based on vanadyl phosphate for fructose dehydration to 5-hydroxymethyl-2-furaldehyde. Appl. Catal. A: General 2004, 275, 111–118.

22. Benvenuti, F.; Carlini, C.; Patrono, P.; Galletti, A.M.R.; Sbrana, G.; Massucci, M.A.; Galli, P. Heterogeneous zirconium and titanium catalysts for the selective synthesis of 5-hydroxymethyl-2- furaldehyde from carbonhydrates. Appl. Catal- A General 2000, 193, 147–153.

23. Román-Leshkov, Y.; Chheda, J.N.; Dumesic, J.A. Phase modifiers promote efficient production of hydroxymethylfurfural from fructose. Science 2006, 312, 1933.

24. Moreau, C.; Durand, R.; Razigade, S.; Duhamet, J. ; Faugeras, P.; Rivalier, P.; Ros, P.; Avignon, G. Dehydration of fructose to 5-hydroxymethylfurfural over H-mordenites. Appl. Catal. A: General 1996, 145, 211–224.

25. Antal, M.J., Jr.; Mok, W.S.L.; Richards, G.N. Mechanism of formation of 5-(hydroxymethyl)-2- furaldehyde from D-fructose and sucrose. Carbohydr. Res. 1990, 199, 91–109.

26. Rosatella, A.A.; Simeonov, S.P.; Frade, R.F.M.; Afonso, C.A.M. 5-Hydroxymeth-ylfurfural (HMF) as a building block platform: Biological properties, synthesis and synthetic applications. Green Chem. 2011, 13, 754–793.

27. Zhao, H.; Holladay, J.E.; Brown, H.; Zhang, Z. Metal chlorides in ionic liquid solvents convert sugars to 5-hydroxymethylfurfural. Science 2007, 316, 1597–1600.

28. Huang, R.; Qi, W.; Su, R.; He, Z. Intergrating enzymatic and acid catalysis to convert glucose into 5-hydroxymethylfurfural. Chem. Commun. 2010, 46, 1115–1117.

29. Khajavi, S.H.; Kimura, Y.; Oomori, T.; Matsuno, R.; Adachi, S. Degradation kinetics of monosaccharides in subcritical water. J. Food Eng. 2005, 68, 309–313.

30. Carlini, C.; Giuttari, M.; Galletti, A.M.R.; Sbrana, G.; Armroli, T.; Busca, G. Selective saccharides dehydration to 5-hydroxymethyl-2-furaldehyde by heterogeneous niobium catalysts. Appl. Catal. A: General 1999, 183, 295–302.

31. Emsley, A.M.; Stevens, G.C. Kinetics and mechanisms of the low-temperature degradation of cellulose. Cellulose 1994, 1, 26–56.

32. Scheiding, W.; Thoma, M.; Ross, A.; Schigerl, K. Modelling of the enzymatic hydrolysis of cellobiose and cellulose by a complex enzyme mixture of Trichoderma reesei QM 9414. Appl. Microbiol. Biotechnol. 1984, 20, 176–182.

33. Shi, N.; Liu, Q.; Wang, T.; Ma, L.; Zhang, Q. High yield production of 5-hydroxymethylfurfural from cellulose by high concentration of sulfates in biphasic system. Green Chem. 2013, Accepted manuscript.

CHAPTER 9

Periodic Mesoporous Organosilica Functionalized with Sulfonic Acid Groups as Acid Catalyst for Glycerol Acetylation

ELS DE CANCK, INMACULADA DOSUNA-RODRÍGUEZ, ERIC M. GAIGNEAUX, AND PASCAL VAN DER VOORT

9.1 INTRODUCTION

The discovery of Periodic Mesoporous Organosilicas (PMOs) [1–3] with organic bridging groups incorporated in their silica framework has been the start of a fascinating research area which provides materials with huge potential [4–6]. Different organic bridges have been employed for very diverse applications, such as heterogeneous catalysts [7,8], bio-sensors [9,10], chromatographic packing materials [11,12], low-k materials [13,14], adsorbents of pollutants [15] and controlled drug delivery systems [16–19]. PMOs are highly porous materials with large specific surface areas, pore volumes and narrow pore size distributions. Furthermore, they exhibit a high thermal and mechanical stability [20–22], especially in comparison with other porous silica materials [23]. This type of material is synthesized with structure directing agents such as the non-ionic triblock co-polymer P123. Around this template, a silica source is condensed in basic or acid aqueous environment. Usually, an organo bis-silane $(R'O)_3$–Si–R–Si–$(OR')_3$ is used where R represents the organic bridging group and R' usually a methyl or ethyl group. Already many reports have appeared on different bridging groups (R) like phenylene, ethylene, ethenylene and ethylbenzene but also more complex and flexible organic functionalities have been de-scribed. Furthermore the bridging group can be modified to fine-tune the material for a specific application such as solid acid catalysis [4].

Concerning this topic, some very promising results have already been published regarding the incorporation of an acid functionality such as a sulfonic acid group and its catalytic activity. Several diverse methods have been applied to prepare sulfonic acid containing PMO materials. These strategies include the direct sulfonation of the phenylene bridge [24–26], as first attempted by Inagaki et al. [27], and the cocondensa-tion of an organo bis-silane with (3-mercaptopropyl) trimethoxysilane (MPTMS) [28–31] followed by an oxidation of the thiol functionality. The latter can also be achieved by the in-situ oxidation of the thiol func-tionality by the addition of H_2O_2 during the cocondensation process of tetraethoxyorthosilicate (TEOS) and MPTMS [32,33]. Other silanes have been used in cocondensation processes with an organo bis-silane such as 2-(4-chlorosulfonylphenyl)-trimethoxysilane [34] and perfluorinated al-kylsulfonic acid silanes [35–37].

In the specific case of ethenylene bridged PMOs, $-SO_3H$ moieties can be acquired by the direct sulfonation of the C=C bond [38]. However, the sulfonic acid group can detach from the material, depending on the environment used during catalysis. Another explored route is the use of a Diels Alder reaction where the ethene bond acts as dienophile. Kondo et al. [39,40] described the cycloaddition of the ethene bond with ben-zocyclobutene and subsequently the resulting phenylene moiety is sulfo-nated to obtain a heterogeneous catalyst. These authors tested this material for several catalytic reactions (esterification of acetic acid with ethanol, the Beckmann and pinacole-pinacolone rearrangement) and the catalyst showed excellent conversion results. This Diels Alder process has been further used to expand the functionalization possibilities [41].

Another example of modifying the surface of an ethenylene bridged PMO material has been reported by the research group of Kaliaguine [42]. First, the surface silanols were end-capped with hexamethyldisilazane after which a Friedel-Crafts alkylation with benzene was made, and subsequently the benzene moiety was sulfonated with concentrated sulfuric acid. The $-SO_3H$ containing material exhibited a high catalytic activity in the self-condensation of heptanal.

In this study, the pure trans-ethenylene bridged PMO material [43] is chosen as support material. Thiol functionalities were incorporated according to a procedure previously described by our research group [44] and subsequently oxidized in order to obtain $-SO_3H$. This material was thoroughly characterized and the solid acid was tested in the acetylation of glycerol. Furthermore, the reusability of this catalyst was investigated.

9.2 RESULTS AND DISCUSSION

Starting from the trans-ethenylene bridged PMO (EP), a sulfonic acid modified PMO material is prepared, characterized and its catalytic activity is studied in the acetylation of glycerol. A general overview of the preparation method, including the starting material, is presented in Figure 1. Firstly, an ethenylene bridged Periodic Mesoporous Organosilica is synthesized starting from E-1,2-bis(triethoxysilyl)ethene (Figure 1, pathway A) [43,45].

Afterwards, the ethene bridge can be further functionalized to incorporate sulfonic acid groups in the material. When pathway B is followed, the material EP is firstly brominated and subsequently the bromine can be substituted via an SN2 reaction by the Grignard reagent of 3-chloro-1-propanethiol (Figure 1, pathway C). A material with a propylthiol group is obtained (EP–$(CH_2)_3$–SH) [44,46]. Oxidizing the thiol moiety is performed with H_2SO_4 or other oxidizing agents such as HNO_3 [28–31] or H_2O_2 [25,47,48]. In this study, a treatment of EP–$(CH_2)_3$–SH with sulfuric acid and a thorough washing step was selected by which it converted the –SH into a –SO_3H group (Figure 1, pathway D). This route resulted in the material EP–$(CH_2)_3$–SO_3H.

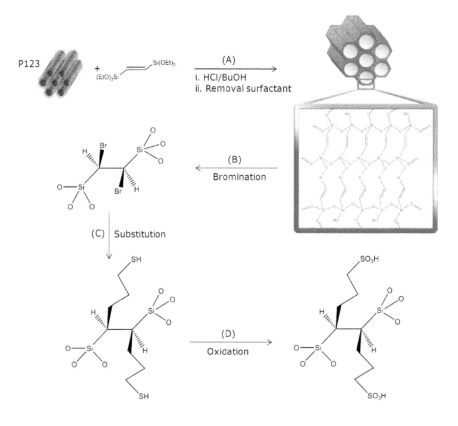

FIGURE 1: Summary of the synthetic pathways followed in this study. (A) Preparation of trans-ethenylene bridged Periodic Mesoporous Organosilica [EP]; (B) Bromination of EP [BEP]; (C) Substitution of the bromine with Grignard reagent of 3-chloro-1-propanethiol [EP–$(CH_2)_3$–SH] and (D) Oxidation with sulfuric acid [EP–$(CH_2)_3$–SO_3H].

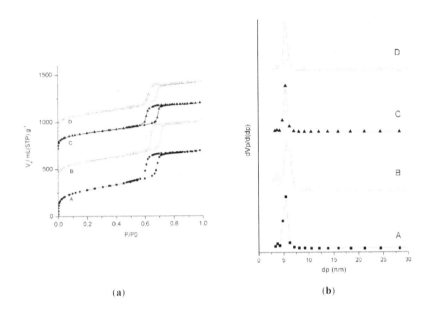

(a)   (b)

FIGURE 2: The nitrogen adsorption and desorption isotherms (a) and pore size distributions (b) of (A) EP; (B) BEP; (C) EP–(CH$_2$)$_3$–SH and (D) EP–(CH$_2$)$_3$–SO$_3$H. The isotherms of BEP, EP–(CH$_2$)$_3$–SH and EP–(CH$_2$)$_3$–SO$_3$H are vertically offset for clarity by 350, 700 and 900 mL (STP) g^{-1} (Standard Temperature and Pressure), respectively.

9.2.1 CHARACTERIZATION OF THE SOLIDS

Nitrogen sorption measurements were performed to examine the porosity of the different materials obtained by the reaction pathway shown in Figure 1 (EP, BEP, EP–(CH$_2$)$_3$–SH and EP–(CH$_2$)$_3$–SO$_3$H). The nitrogen adsorption and desorption isotherms are shown in Figure 2. The type IV isotherms with the condensation step at relative pressures between 0.55 and 0.75 and the H1 hysteresis of the solids clearly indicate that the materials are mesoporous and possess cylindrical pores with a narrow pore size distribution.

A summary of the properties of these materials is shown in Table 1. The materials exhibit high specific surface areas (SBET) ranging from 850 to 523 m² g⁻¹ and large total pore volumes around 0.84 mL g⁻¹. The SBET decreases when the material is functionalized due to the decoration of the pore walls with the bromine and later on with the propylthiol functionality but also due to the overall weight gain of the functionalized materials. The pore diameter of all the materials lies in the range of 6 to 5 nm. Only a minor shift to smaller pore diameters and a slight broadening of the pore size distribution is observed (Figure 2 and Table 1). The structural characteristics of the commercially available resin Amberlyst-15 are also presented in Table 1 for comparison. This ethenylbenzenesulfonic acid polymer is a strong acid ion exchange resin with unordered macropores. The material is also prone to swelling.

TABLE 1: Overview of the structural characteristics of the materials compared in this study.

Sample	Path	SBET^a (m$_2$ g^{-1})	Vp^b (mL g^{-1})	dp^c (nm)
EP	A	850	1.03	5.8
BEP	B	663	0.84	5.6
EP–(CH$_2$)$_3$–SH	C	523	0.59	5.3
EP–(CH$_2$)$_3$–SO$_3$H	D	688	0.72	5.4
Amberlyst-15	–	50	–	300

Notes: ^{a}Surface area calculated via the Brunauer-Emmett-Teller (BET) model; ^{b}Total pore volume at P/P0 = 0.98; ^{c}Pore diameter calculated via the Barrett-Joyner-Halenda (BJH) plot.

The XRD patterns of the materials in Figure 3 reveal three well-resolved signals originating from the low angle (100) and second-order (110) and (200) reflections. This evidently indicates that the materials possess a 2D-hexagonal ordered structure and thus retain their P6mm space group ordering throughout the syntheses. Only a slight broadening can be observed at the patterns of sample BEP, EP–(CH$_2$)$_3$–SH and EP–(CH$_2$)$_3$–SO$_3$H.

It is quite remarkable that all the materials discussed in this study, show outstanding structural stability. The materials retain porosity and ordering after three consecutive reactions as can be seen from the nitrogen sorption and XRD data. These results also confirm the reported stability of Periodic Mesoporous Organosilicas [20,23].

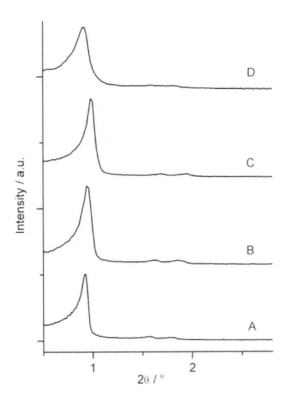

FIGURE 3: The powder X-ray diffraction patterns of (A) EP; (B) BEP; (C) EP–(CH$_2$)$_3$–SH; and (D) EP–(CH$_2$)$_3$–SO$_3$H.

Table 2 presents an overview of the chemical characterization of the solids after the different synthetic procedures. The bromination of the ethene bridge is a very straightforward reaction and approximately 25%–30% of the double bonds are brominated as the remaining fraction of double bonds are buried inside the walls and are unavailable for further reaction [45]. The subsequent substitution with the Grignard reagent results in thiol functionalities. The amount of thiol functionalities is determined using a silver titration [44,46]. After oxidation of the thiol groups using sulfuric acid and a thorough washing step, a total amount of 0.60 mmol H$^+$ per gram of material has been observed. This also includes the intrinsic acidity of the PMO material originating from the surface silanols (~0.15 mmol g^{-1}), as we described earlier [49]. It is clear that the conversion of the thiol containing PMO into the sulfonic acid containing-material has occurred via the oxidation process. This is also confirmed by Raman spectroscopy by the appearance of two signals in the region between 1160 and 1190 cm^{-1} (See figure S1 in supplementary information). Also, the thiol titration after oxidation showed a zero concentration of remaining thiol groups. Amberlyst-15 exhibits a high acidity of 4.7 mmol H^+ g^{-1}.

TABLE 2: Overview of the chemical characteristics of the materials compared in this study.

Sample	Functionality	mmol g^{-1}
BEP	$-Br^a$	2.39
EP–$(CH_2)_3$–SH	$-SH^b$	0.40
EP–$(CH_2)_3$–SO_3H	$-SO_3H^c$	0.60^d

Notes: *a*Determined gravimetrically; *b*Determined via silver titration; *c*Determined via acid/base titration; d The deviation between the amount of thiols and total acidity is due to the acidity of the surface silanols.

9.2.2 CATALYTIC EXPERIMENTS AND RECYCLABILITY

The catalytic ability of the sulfonic acid functionalized PMO material has been explored for an esterification reaction, i.e., the glycerol acetylation reaction (Figure 4). The activity of EP–(CH2)3–SO3H is compared with a commercially available catalyst Amberlyst-15 and moreover the catalysts' reusability is explored.

In this study the esterification of glycerol is probed due to its economic importance. Glycerol is an important by-product of first generation biodiesel and is produced in a relative large quantity [50]. This overproduction of glycerol can be used in order to develop second generation biodiesel which uses glycerol as a raw product. As carboxylic acid, acetic acid is probed as shown in the general reaction (Figure 4). Three products may in principle be obtained from this reaction: glycerol monoacetate (MAG),

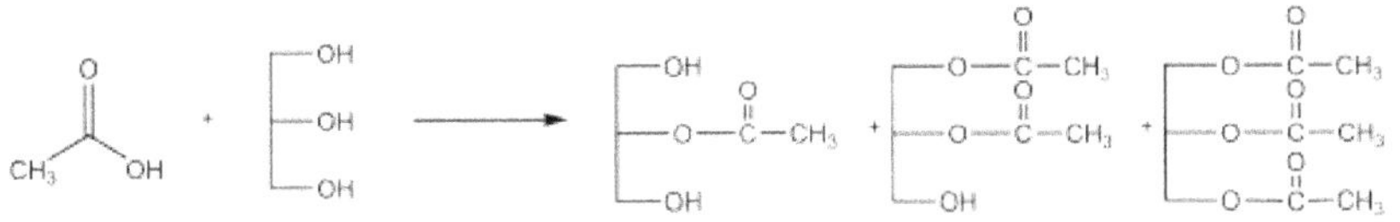

FIGURE 4: The esterification reaction: the acetylation of glycerol with the formation of glycerol monoacetate, glycerol diacetate and glycerol triacetate.

glycerol diacetate (DAG) and glycerol triacetate (TAG). However, experimentally, only MAG (~94%) and DAG (~6%) are formed using the specific catalytic conditions described in the experimental part [51].

The catalytic activity of EP–$(CH_2)_3$–SO_3H for the esterification of acetic acid with glycerol is presented in Figure 5 where the total acetylation yield is shown as a function of time. The total acetylation yield is defined according to the equation below:

$$Yield_\tau \, (\%) = ([P]_t/[HAC]_0) \, (\upsilon_{HAC}/\upsilon_p) \times 100 \tag{1}$$

where $[P]_t$ and $[HAc]_0$ represent the product and acetic acid concentration at a certain reaction time and at t = 0, respectively. Furthermore, υ_{HAc} and υ_p represent the stoichiometric coefficients of the acetic acid and the ester formed, i.e., 1 for mono-substituted, 2 for di-substituted and 3 for fully substituted products, respectively. Also, as acetic acid contains acid protons which can induce a self-catalyzed process, the reaction in absence of any solid catalyst was monitored. Corresponding data are shown in Figure 5. It is clear that the sulfonated PMO possesses a significant catalytic activity with a yield of almost 80% for this reaction after ~300 min; whereas the blank test (without any solid catalyst involved) yielded only ~50% of esters after ~300 min. The conversion of Amberlyst-15 is shown in the same figure. Comparing the two materials clearly shows that EP–$(CH_2)_3$–SO_3H exhibits a similar catalytic activity as Amberlyst-15, which is a well-performing catalyst in this type of reaction.

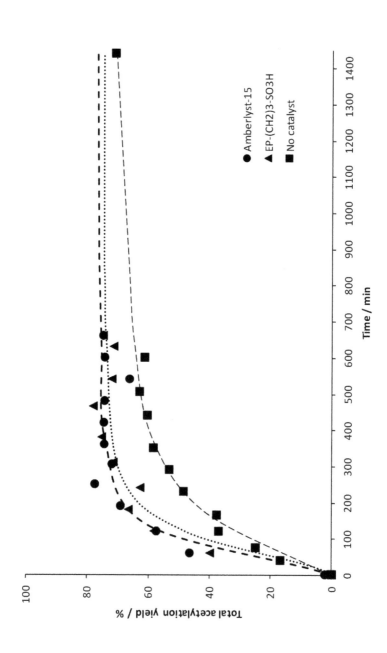

FIGURE 5: The total acetylation yield for the catalytic reaction with EP–(CH$_2$)$_3$–SO$_3$H and Amberlyst-15. Also the blank reaction is represented for clarity. A catalyst loading of 0.25 g per 40 mL of glycerol was used. The lines are intended as visual aids only.

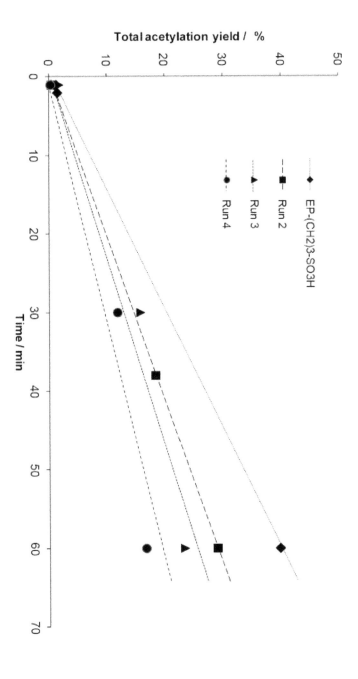

FIGURE 6: Recyclability experiments for EP–(CH$_2$)$_3$–SO$_3$H with several runs during the first hour of the reaction. The lines are intended as visual aids only.

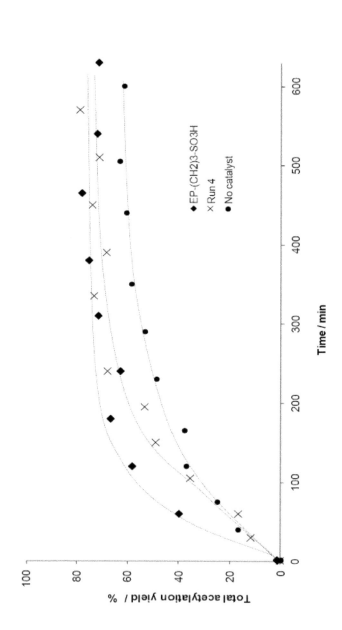

FIGURE 7: Recyclability experiment for EP–(CH$_2$)$_3$–SO$_3$H: a comparison between the catalytic activity of the pristine material and the third catalytic run. The blank reaction is presented for clarity. The lines are intended as visual aids only.

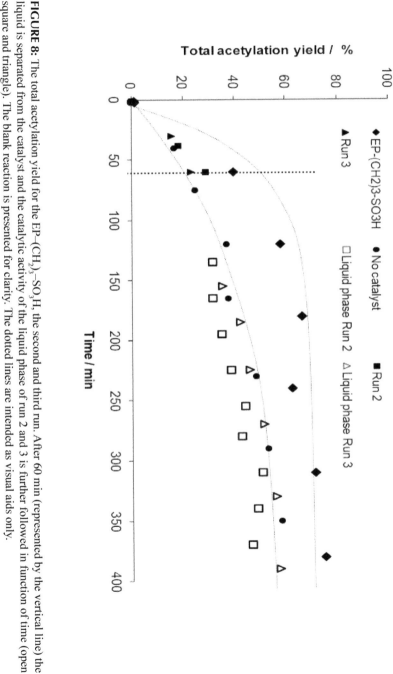

FIGURE 8: The total acetylation yield for the EP–(CH$_2$)$_3$–SO$_3$H, the second and third run. After 60 min (represented by the vertical line) the liquid is separated from the catalyst and the catalytic activity of the liquid phase of run 2 and 3 is further followed in function of time (open square and triangle). The blank reaction is presented for clarity. The dotted lines are intended as visual aids only.

Furthermore, the recyclability of the sulfonic acid containing PMO material is studied for three consecutive runs. First, the initial rate of the catalytic reaction is studied for each run by focusing on the first hour of the acetylation (Figure 6). These experiments are all performed in the same catalytic set-up as the standard catalytic experiment. The solid is filtered after 1 hour and re-used without any further treatment in the subsequent run with a fresh reaction medium (run 2); this being performed again for two additional consecutive runs (runs 3 and 4). As one can see from the figure, the material still possesses catalytic activity for the acetylation after four runs. However, recycling of the material in the consecutive runs results in a slight decrease of the initial reaction rate.

After the last catalytic run, i.e., run 4, the total acetylation yield is monitored for approximately 10 h in order to compare it with the acetylation yield of the pristine sulfonated PMO material, EP–$(CH_2)_3$–SO_3H (Figure 7). Although a decrease in the initial reaction rate was observed as already mentioned, at the third consecutive run, the material still reaches the equilibrium after approximately 5 h and finally results in the same acetylation level as the fresh pristine material.

Moreover, additional tests are performed to evaluate the actual heterogeneous character of the observed catalytic activity. Therefore, the solids are removed from the liquid after 1 h along the first and second runs, and the corresponding recovered solutions are kept under the catalytic conditions to follow the occurrence of a further evolution of the total acetylation yield without solid catalyst in the system anymore. As can be seen from Figure 8, it is clear that the acetylation occurs much slower, i.e., in the range of the blank, when the catalysts are removed of the reaction media, than when the catalyst is maintained in the reactor for the whole test duration.

9.3 EXPERIMENTAL SECTION

9.3.1 MATERIALS

The following reagents were used for the synthesis and characterization: Grubbs' first generation catalyst $(PCy_3)Cl_2Ru = CH–Ph$, pluronic P123 $(PEO_{20}PPO_{70}PEO_{20})$, 1-butanol, bromine, 3-chloro-1-propanethiol (98%), magnesium turnings (98%), sodium chloride, potassium thiocyanate, ammonium iron sulfate dodecahydrate, nitric acid (65%), iron (III) chloride (98%; anhydrous), toluene (anhydrous), acetonitrile, sulfuric acid (H_2SO_4), silver nitrate were obtained from Sigma-Aldrich (Bornem, Belgium). Acetone (>99.5%) was acquired from VWR (Belgium). Tetrahydrofurane (rotidry) and hydrochloric acid (37%; p.a.) were purchased at Carl Roth (Karlsruhe, Germany). Vinyltriethoxysilane (97%) was acquired from ABCR (Karlsruhe, Germany). The following reagents were used for the catalytic reaction: acetic acid (puriss. p.a., ACS reagens, ≥99.8%, GC/T), glycerol (puriss. p.a., ACS reagens, anhydrous, dist., ≥99.5%) were purchased from Fluka (Bornem, Belgium). 1-Propanol, 1-heptanol (98%) and o-xylene (98%) were obtained from Sigma Aldrich. All chemicals were used as received.

9.3.2 SYNTHESIS OF PURE E-1,2-BIS(TRIETHOXYSILYL)ETHENE (E-BTSE)

The diastereoisomerically pure E-isomer of the ethenylene precursor was synthesized according to a procedure described by our research group [43,45]. An amount of 0.0535 g of the Grubbs' first generation catalyst and 42.95 mL of vinyltriethoxysilane (VTES) were mixed together under inert atmosphere. The mixture was stirred for 1 h at room temperature and subsequently refluxed. After 3 h, a distillation was performed to remove remaining VTES. Afterwards, the colorless E-1,2-bis(triethoxysilyl)ethene (E-BTSE) was distilled off under vacuum (~0.1 Pa).

9.3.3 SYNTHESIS OF ETHENYLENE BRIDGED PERIODIC MESOPOROUS ORGANOSILICA (FIGURE 1; PATHWAY A)

In a typical synthesis, reported by our group [45], a total amount of 1.00 g of P123 was placed in a flask and 47.80 mL of distilled H_2O, 3.42 mL concentrated HCl and 2.45 mL 1-butanol were added. The solution was stirred for 1.5 h at room temperature and subsequently 1.86 mL of E-BTSE was added to the mixture. Next, the mixture was vigorously stirred at 35 °C. After 4 h, the temperature was raised to 90 °C and aged for 16 h. Afterwards, the mixture was left to cool down, and the white solid was filtered and washed with acetone and distilled water. Finally, P123 was removed via Soxhlet extraction with acetone for 5 h. The white solid, denoted as EP, was dried at 120 °C under vacuum (~0.1 Pa).

9.3.4 SYNTHESIS OF BEP AND EP-PROPYLTHIOL (FIGURE 1; PATHWAYS B AND C)

The synthesis of this material was extensively described by our research group [38,46]. A certain amount of EP was first brominated with bromine gas under vacuum for 3 h. The resulting material was then dried at 90 °C for 16 h. Magnesium (0.74 g), iron(III) chloride (0.54 g) and dry tetrahydrofuran (THF) (30 mL) were mixed under inert atmosphere and stirred for 30 min at 50 °C. Next, 3-chloro-1-propanethiol (0.22 mL) was added and left to stir for 4 h at room temperature. Subsequently, the solution was added to 0.70 g of dry brominated EP. The resulting mixture was stirred for an additional 24 h at 40 °C. Then, the solid was filtered and washed several times with THF, 2 mol L^{-1} HCl, H$_2$O and acetone. Finally, the material was dried at 90 °C for 16 h under vacuum (~0.1 Pa) and denoted as EP–(CH$_2$)$_3$–SH.

9.3.5 SYNTHESIS OF EP-PROPYLSULFONIC ACID (FIGURE 1; PATHWAY D)

A volume of 25 mL H$_2$SO$_4$ (2.5 mol L^{-1}) was added to 0.50 g of EP–(CH$_2$)$_3$–SH and stirred at room temperature. After 1 h, the solids were filtered and washed thoroughly with water and acetone. Finally, the material was dried at 90 °C for 16 h under vacuum (~0.1 Pa) and referred to as EP–(CH$_2$)$_3$–SO$_3$H.

9.3.6 DETERMINATION OF THE AMOUNT OF REACHABLE THIOLS

A titration is performed as described in the literature [44]. A solution of silver nitrate with a known concentration is added to 0.050 g of material and left to stir until the equilibrium is reached. The excess of silver is titrated with potassium thiocyanate and $FeNH_4(SO_4)_2 \cdot 12H_2O$ in 0.3 mol L^{-1} HNO_3 as indicator.

9.3.7 DETERMINATION OF THE TOTAL ACIDITY

The material containing sulfonic acid groups was stirred for 24 h in 20 g of 2 mol L^{-1} NaCl. Next, an acid-base titration was achieved with NaOH and phenolphtaleine as indicator.

9.3.8 ESTERIFICATION OF ACETIC ACID WITH GLYCEROL

The catalytic reaction was carried out at 105 °C and at atmospheric pressure in a round bottom flask reactor equipped with a magnetic stirrer. The sulfonic acid containing material was dried prior to use at 105 °C under vacuum (~1.33 Pa) for 18 h. The following concentrations of the reagents were used: 100 g acetic acid/L of glycerol and the catalyst concentration was 6.25 g/L of glycerol. First, the catalyst and glycerol are added together to the flask and are brought to reaction temperature. When the temperature is reached, the acetic acid is injected. Before analysis the samples are extracted with 1-heptanol and using o-xylene as an internal standard. The total yield of the esters was determined with gas chromatography. After one minute a sample of the reaction mixture is taken and analyzed with gas chromatography. This sample represents the starting point of the catalytic experiment. The recyclability experiments were carried out under identical conditions as the catalytic tests. Therefore, the catalyst was separated from the liquid phase after 1 h and reused in a consecutive run without further treatment. The liquid phase was kept under reaction conditions and further followed as a function of time. Gas chromatography was performed with a Varian CP-3800 using a Cp-Sil 8CD column (Varian, Palo Alto, CA, USA) with a FID detector.

9.3.9 CHARACTERIZATION

Nitrogen sorption measurements were carried out by a Belsorp-mini II gas analyzer at 77 K. The specific surface area (SBET) was determined by the BET equation (p/p0 = 0.05–0.15). The pore size distribution was determined from the desorption branch of the isotherm using the Barrett-Joyner-Halenda (BJH) theory. The samples were pretreated at 90 °C while degassing (~0.1 Pa). X-ray diffraction (XRD) measurements were conducted with an ARL X'tra X-ray diffractometer of Thermo Scientific (Waltham, MA, USA) equipped with a Cu Kα1 tube and a Peltier cooled lithium drifted silicon solid stage detector. Raman spectra are recorded using a Raman type FRA106/S spectrometer of Bruker (Karlsruhe, Germany), equipped with a Nd-YAG laser (λ = 1064 nm).

9.4 CONCLUSIONS

Periodic Mesoporous Organosilica functionalized with sulfonic acid groups has been successfully synthesized and characterized. An ethenylene bridged PMO material was chosen as starting material and this was further post-modified in several steps. A bromination and subsequent substitution reaction was used followed by an oxidation turning the material into a solid acid catalyst. This material has been investigated for the acetylation of glycerol. The material showed an equal activity as Amberlyst-15; the latter being usually considered as the reference most-efficient catalyst material for such kind of reactions. Recyclability experiments showed that the sulfonated Periodic Mesoporous Organosilica is reusable and showed the same total acetylation yield after three runs. Homogeneous tests showed that the acetylation of glycerol indeed occurs heterogeneously, suggesting the absence of leaching, e.g., of sulfonate species, in the medium, and thus the high stability of our material under working conditions.

REFERENCES

1. Inagaki, S.; Guan, S.; Fukushima, Y.; Ohsuna, T.; Terasaki, O. Novel mesoporous materials with a uniform distribution of organic groups and inorganic oxide in their frameworks. J. Am. Chem. Soc. 1999, 121, 9611–9614.
2. Asefa, T.; MacLachan, M.J.; Coombs, N.; Ozin, G.A. Periodic mesoporous organosilicas with organic groups inside the channel walls. Nature 1999, 402, 867–871.

3. Melde, B.J.; Holland, B.T.; Blanford, C.F.; Stein, A. Mesoporous sieves with unified hybrid inorganic/organic frameworks. Chem. Mater. 1999, 11, 3302–3308.
4. Van der Voort, P.; Esquivel, D.; de Canck, E.; Goethals, F.; van Driessche, I.; Romero-Salguero, F.J. Periodic Mesoporous Organosilicas: from simple to complex bridges; A comprehensive overview of functions, morphologies and applications. Chem. Soc. Rev. 2013, 42, 3913–3955.
5. Mizoshita, N.; Tani, T.; Inagaki, S. Syntheses, properties and applications of periodic mesoporous organosilicas prepared from bridged organosilane precursors. Chem. Soc. Rev. 2011, 40, 789–800.
6. Hoffmann, F.; Froba, M. Vitalising porous inorganic silica networks with organic functions-PMOs and related hybrid materials. Chem. Soc. Rev. 2011, 40, 608–620.
7. Sasidharan, M.; Fujita, S.; Ohashi, M.; Goto, Y.; Nakashima, K.; Inagaki, S. Novel synthesis of bifunctional catalysts with different microenvironments. Chem. Commun. 2011, 47, 10422–10424.
8. Ohashi, M.; Kapoor, M.P.; Inagaki, S. Chemical modification of crystal-like mesoporous phenylene-silica with amino group. Chem. Commun. 2008, 7, 841–843.
9. Johnson-White, B.; Zeinali, M.; Shaffer, K.M.; Patterson, C.H.; Charles, P.T.; Markowitz, M.A. Detection of organics using porphyrin embedded nanoporous organosilicas. Biosens. Bioelectron. 2007, 22, 1154–1162.
10. Johnson-White, B.; Zeinali, M.; Malanoski, A.P.; Dinderman, M.A. Sunlight-catalyzed conversion of cyclic organics with novel mesoporous organosilicas. Catal. Commun. 2007, 8, 1052–1056.
11. Huang, L.L.; Lu, J.; Di, B.; Feng, F.; Su, M.X.; Yan, F. Self-assembled highly ordered ethane-bridged periodic mesoporous organosilica and its application in HPLC. J. Sep. Sci. 2011, 34, 2523–2527.
12. Zhang, Y.P.; Jin, Y.; Yu, H.; Dai, P.C.; Ke, Y.X.; Liang, X.M. Pore expansion of highly monodisperse phenylene-bridged organosilica spheres for chromatographic application. Talanta 2010, 81, 824–830.
13. Goethals, F.; Baklanov, M.R.; Ciofi, I.; Detavernier, C.; van der Voort, P.; van Driessche, I. A new procedure to seal the pores of mesoporous low-k films with precondensed organosilica oligomers. Chem. Commun. 2012, 48, 2797–2799.
14. Goethals, F.; Ciofi, I.; Madia, O.; Vanstreels, K.; Baklanov, M.R.; Detavernier, C.; van der Voort, P.; van Driessche, I. Ultra-low-k cyclic carbon-bridged PMO films with a high chemical resistance. J. Mater. Chem. 2012, 22, 8281–8286.
15. Walcarius, A.; Mercier, L. Mesoporous organosilica adsorbents: Nanoengineered materials for removal of organic and inorganic pollutants. J. Mater. Chem. 2010, 20, 4478–4511.
16. Parambadath, S.; Rana, V.K.; Moorthy, S.; Chu, S.W.; Park, S.K.; Lee, D.; Sung, G.; Ha, C.S. Periodic mesoporous organosilicas with co-existence of diurea and sulfanilamide as an effective drug delivery carrier. J. Solid State Chem. 2011, 184, 1208–1215.
17. Parambadath, S.; Rana, V.K.; Zhao, D.Y.; Ha, C.S. N,N'-diureylenepiperazine-bridged periodic mesoporous organosilica for controlled drug delivery. Microporous Mesoporous Mater. 2011, 141, 94–101.
18. Vathyam, R.; Wondimu, E.; Das, S.; Zhang, C.; Hayes, S.; Tao, Z.M.; Asefa, T. Improving the adsorption and release capacity of organic-functionalized mesoporous

materials to drug molecules with temperature and synthetic methods. J. Phys. Chem. C 2011, 115, 13135–13150.

19. Moorthy, M.S.; Park, S.S.; Fuping, D.; Hong, S.H.; Selvaraj, M.; Ha, C.S. Step-up synthesis of amidoxime-functionalised periodic mesoporous organosilicas with an amphoteric ligand in the framework for drug delivery. J. Mater. Chem. 2012, 22, 9100–9108.

20. Goethals, F.; Vercaemst, C.; Cloet, V.; Hoste, S.; van der Voort, P.; van Driessche, I. Comparative study of ethylene- and ethenylene-bridged periodic mesoporous organosilicas. Microporous Mesoporous Mater. 2010, 131, 68–74.

21. Kruk, M.; Jaroniec, M.; Guan, S.Y.; Inagaki, S. Adsorption and thermogravimetric characterization of mesoporous materials with uniform organic-inorganic frameworks. J. Phys. Chem. B 2001, 105, 681–689.

22. Esquivel, D.; Jimenez-Sanchidrian, C.; Romero-Salguero, F.J. Comparison of the thermal and hydrothermal stabilities of ethylene, ethylidene, phenylene and biphenylene bridged periodic mesoporous organosilicas. Mater. Lett. 2011, 65, 1460–1462.

23. Burleigh, M.C.; Markowitz, M.A.; Jayasundera, S.; Spector, M.S.; Thomas, C.W.; Gaber, B.P. Mechanical and hydrothermal stabilities of aged periodic mesoporous organosilicas. J. Phys. Chem. B 2003, 107, 12628–12634.

24. Esquivel, D.; Jimenez-Sanchidrian, C.; Romero-Salguero, F.J. Thermal behaviour, sulfonation and catalytic activity of phenylene-bridged periodic mesoporous organosilicas. J. Mater. Chem. 2011, 21, 724–733.

25. Rac, B.; Hegyes, P.; Forgo, P.; Molnar, A. Sulfonic acid-functionalized phenylene-bridged periodic mesoporous organosilicas as catalyst materials. Appl. Catal. A Gen. 2006, 299, 193–201.

26. Kapoor, M.P.; Kasama, Y.; Yanagi, M.; Yokoyama, T.; Inagaki, S.; Shimada, T.; Nanbu, H.; Juneja, L.R. Cubic phenylene bridged mesoporous hybrids from allylorganosilane precursors and their applications in Friedel-Crafts acylation reaction. Microporous Mesoporous Mater. 2007, 101, 231–239.

27. Inagaki, S.; Guan, S.; Ohsuna, T.; Terasaki, O. An ordered mesoporous organosilica hybrid material with a crystal-like wall structure. Nature 2002, 416, 304–307.

28. Yang, Q.H.; Kapoor, M.P.; Shirokura, N.; Ohashi, M.; Inagaki, S.; Kondo, J.N.; Domen, K. Ethane-bridged hybrid mesoporous functionalized organosilicas with terminal sulfonic groups and their catalytic applications. J. Mater. Chem. 2005, 15, 666–673.

29. Yang, Q.H.; Kapoor, M.P.; Inagaki, S.; Shirokura, N.; Kondo, J.N.; Domen, K. Catalytic application of sulfonic acid functionalized mesoporous benzene-silica with crystal-like pore wall structure in esterification. J. Mol. Catal. A Chem. 2005, 230, 85–89.

30. Yang, Q.H.; Liu, J.; Yang, J.; Kapoor, M.P.; Inagaki, S.; Li, C. Synthesis, characterization, and catalytic activity of sulfonic acid-functionalized periodic mesoporous organosilicas. J. Catal. 2004, 228, 265–272.

31. Yang, Q.H.; Kapoor, M.P.; Inagaki, S. Sulfuric acid-functionalized mesoporous benzene-silica with a molecular-scale periodicity in the walls. J. Am. Chem. Soc. 2002, 124, 9694–9695.

32. Melero, J.A.; Vicente, G.; Morales, G.; Paniagua, M.; Moreno, J.M.; Roldan, R.; Ezquerro, A.; Perez, C. Acid-catalyzed etherification of bio-glycerol and isobutylene over sulfonic mesostructured silicas. Appl. Catal. A Gen. 2008, 346, 44–51.

33. Melero, J.A.; Bautista, L.F.; Morales, G.; Iglesias, J.; Sanchez-Vazquez, R. Biodiesel production from crude palm oil using sulfonic acid-modified mesostructured catalysts. Chem. Eng. J. 2010, 161, 323–331.

34. Rat, M.; Zahedi-Niaki, M.H.; Kaliaguine, S.; Do, T.O. Sulfonic acid functionalized periodic mesoporous organosilicas as acetalization catalysts. Microporous Mesoporous Mater. 2008, 112, 26–31.

35. Dube, D.; Rat, M.; Shen, W.; Nohair, B.; Beland, F.; Kaliaguine, S. Perfluorinated alkylsulfonic acid functionalized periodic mesostructured organosilica: A new acidic catalyst. Appl. Catal. A Gen. 2009, 358, 232–239.

36. Dube, D.; Rat, M.; Shen, W.; Beland, F.; Kaliaguine, S. Perfluoroalkylsulfonic acid-functionalized periodic mesostructured organosilica: a strongly acidic heterogeneous catalyst. J. Mater. Sci. 2009, 44, 6683–6692.

37. Shen, W.; Dube, D.; Kaliaguine, S. Alkylation of isobutane/1-butene over periodic mesoporous organosilica functionalized with perfluoroalkylsulfonic acid group. Catal. Commun. 2008, 10, 291–294.

38. De Canck, E.; Vercaemst, C.; Verpoort, F.; van der Voort, P. A New Sulphonic Acid Functionalized Periodic Mesoporous Organosilica as A Suitable Catalyst. In Scientific Bases for the Preparation of Heterogeneous Catalysts: Proceedings of the 10th International Symposium; Gaigneaux, E.M.; Devillers, M.; Hermans, S.; Jacobs, P.A.; Martens, J.A.; Ruiz, P., Eds.; Elsevier Science BV: Amsterdam, The Netherlands, 2010; Volume 175, pp. 365–368.

39. Nakajima, K.; Tomita, I.; Hara, M.; Hayashi, S.; Domen, K.; Kondo, J.N. A stable and highly active hybrid mesoporous solid acid catalyst. Adv. Mater. 2005, 17, 1839–1842.

40. Nakajima, K.; Tomita, I.; Hara, M.; Hayashi, S.; Domen, K.; Kondo, J.N. Development of highly active SO3H-modified hybrid mesoporous catalyst. Catal. Today 2006, 116, 151–156.

41. Esquivel, D.; de Canck, E.; Jimenez-Sanchidrian, C.; van der Voort, P.; Romero-Salguero, F.J. Formation and functionalization of surface Diels-Alder adducts on ethenylene-bridged periodic mesoporous organosilica. J. Mater. Chem. 2011, 21, 10990–10998.

42. Dube, D.; Rat, M.; Beland, F.; Kaliaguine, S. Sulfonic acid functionalized periodic mesostructured organosilica as heterogeneous catalyst. Microporous Mesoporous Mater. 2008, 111, 596–603.

43. Vercaemst, C.; Ide, M.; Allaert, B.; Ledoux, N.; Verpoort, F.; van der Voort, P. Ultrafast hydrothermal synthesis of diastereoselective pure ethenylene-bridged periodic mesoporous organosilicas. Chem. Commun. 2007, 22, 2261–2263.

44. De Canck, E.; Lapeire, L.; de Clercq, J.; Verpoort, F.; van der Voort, P. New ultrastable mesoporous adsorbent for the removal of mercury ions. Langmuir 2010, 26, 10076–10083.

45. Vercaemst, C.; Ide, M.; Wiper, P.V.; Jones, J.T.A.; Khimyak, Y.Z.; Verpoort, F.; van ver Voort, P. Ethenylene-Bridged Periodic Mesoporous Organosilicas: From E to Z. Chem. Mater. 2009, 21, 5792–5800.

46. Gao, Y.; de Canck, E.; Leermakers, M.; Baeyens, W.; van der Voort, P. Synthesized mercaptopropyl nanoporous resins in DGT probes for determining dissolved mercury concentrations. Talanta 2011, 87, 262–267.

47. Hamoudi, S.; Royer, S.; Kaliaguine, S. Propyl- and arene-sulfonic acid functionalized periodic mesoporous organosilicas. Microporous Mesoporous Mater. 2004, 71, 17–25.

48. Hamoudi, S.; Kaliaguine, S. Sulfonic acid-functionalized periodic mesoporous organosilica. Microporous Mesoporous Mater. 2003, 59, 195–204.

49. Ide, M.; El-Roz, M.; de Canck, E.; Vicente, A.; Planckaert, T.; Bogaerts, T.; van Driessche, I.; Lynen, F.; van Speybroeck, V.; Thybault-Starzyk, F.; van der Voort, P. Quantification of silanol sites for the most common mesoporous ordered silicas and organosilicas: total versus accessible silanols. Phys. Chem. Chem. Phys. 2013, 15, 642–650.

50. Nigam, P.S.; Singh, A. Production of liquid biofuels from renewable resources. Prog. Energy Combust. Sci. 2011, 37, 52–68.

51. Dosuna-Rodriguez, I.; Gaigneaux, E.M. Glycerol acetylation catalysed by ion exchange resins. Catal. Today 2012, 195, 14–21.

PART III

OPTIMIZING CATALYTIC PERFORMANCE

CHAPTER 10

Alkaline and Alkaline-Earth Ceramic Oxides for CO_2 Capture, Separation and Subsequent Catalytic Chemical Conversion

MARGARITA J. RAMÍREZ-MORENO, ISSIS C. ROMERO-IBARRA, JOSÉ ORTIZ-LANDEROS, AND HERIBERTO PFEIFFER

10.1 INTRODUCTION

The amounts of anthropogenic carbon dioxide (CO_2) in the atmosphere have been raised dramatically, mainly due to the combustion of different carbonaceous materials used in energy production, transport and other important industries such as cement production, iron and steelmaking. To solve or at least mitigate this environmental problem, several alternatives have been proposed. A promising alternative for reducing the CO_2 emissions is the separation and/or capture and concentration of the gas and its subsequent chemical transformation. In that sense, a variety of materials have been tested containing alkaline and/or alkaline-earth oxide ceramics and have been found to be good options.

Margarita J. Ramírez-Moreno, Issis C. Romero-Ibarra, José Ortiz-Landeros and Heriberto Pfeiffer (2014). Alkaline and Alkaline-Earth Ceramic Oxides for CO₂ Capture, Separation and Subsequent Catalytic Chemical Conversion, CO₂ Sequestration and Valorization, *Victor Esteves (Ed.), ISBN: 978-953-51-1225-9, InTech, DOI: 10.5772/57444.*

The aforementioned ceramics are able to selectively trap CO_2 under different conditions of temperature, pressure, humidity and gas mixture composition. The influence of those factors on the CO_2 capture (physically or chemically) seems to promote different sorption mechanisms, which depend on the material's chemical composition and the sorption conditions used. Actually, this capture performance suggests the feasibility of these kinds of solid for being used with different capture technologies and processes, such as: pressure swing adsorption (PSA), vacuum swing adsorption (VSP), temperature swing adsorption (TSA) and water gas shift reaction (WGSR). Therefore, the fundamental study regarding this matter can help to elucidate the whole phenomena in order to enhance the sorbents' properties.

10.2 CO_2 CAPTURE BY DIFFERENT ALKALINE AND ALKALINE-EARTH CERAMICS

Among the alkaline and/or alkaline-earth oxides, various lithium, sodium, potassium, calcium and magnesium ceramics have been proposed for CO_2 capture through adsorption and chemisorption processes [1-20]. These materials can be classified into two large groups: dense and porous ceramics. Dense ceramics mainly trap CO_2 chemically: the CO_2 is chemisorbed. Among these ceramics, CaO is the most studied one. It presents very interesting sorption capacities at high temperatures ($T \geq 600$ °C). In addition to this material, alkaline ceramic oxides have been considered as possible captors, mostly lithium and sodium based ceramics (Li_5AlO_4 and Na_2ZrO_3, for example). In these cases, one of the most interesting properties is related to the wide temperature range in which some of these ceramics trap CO_2 (between 150 and 800 °C), as well as their high CO_2 capture capacity.

In these ceramics, the CO_2 capture occurs chemically, through a chemisorption process. At a micrometric scale, a general reaction mechanism has been proposed, where the following steps have been established: Initially, CO_2 reacts at the surface of the particles, producing the respective alkaline or alkaline-earth carbonate and in some cases different secondary phases. Some examples are:

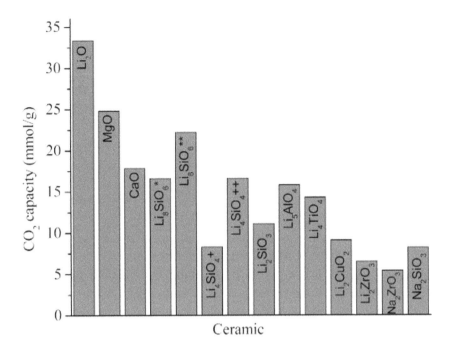

FIGURE 1: Theoretical CO_2 capture capacities for different alkaline and alkaline-earth ceramics. In the Li_8SiO_6 (labeled as *) and Li_4SiO_4 (labeled as +), the maximum capacity can depend on the CO_2 moles captured in each different phase formed ($Li_8SiO_6 + CO_2 \rightarrow Li_4SiO_4 + CO_2 \rightarrow Li_2SiO_3 + Li_2CO_3$).

$$Li_5AlO_4 + 2CO_2 \rightarrow 2Li_2CO_3 + LiAlO_2 \tag{1}$$

$$Na_2ZrO_3 + CO_2 \rightarrow Na_2CO_3 + ZrO_2 \tag{2}$$

$$CaO + CO_2 \rightarrow CaCO_3 \tag{3}$$

The above reactions show that surface products can be composed of carbonates, but as well they can contain metal oxides or other alkaline/alkaline-earth ceramics. The presence of these secondary phases can modify (improve or reduce) the diffusion processes described below [1].

Once the external carbonate shell is formed, different diffusion mechanisms have to be activated in order to continue the CO_2 chemisorption, through the particle bulk. Some of the diffusion processes correspond to the CO_2 diffusion through the mesoporous external carbonate shell, and some others such as the intercrystalline and grain boundary diffusion processes [1, 18, 21].

Figure 1 shows the theoretical CO_2 chemisorption capacities (mmol of CO_2 per gram of ceramic) for the most studied alkaline and alkaline-earth ceramics. As it can be seen, metal oxides (Li_2O, MgO and CaO) are among the materials with the best CO_2 capture capacities. Nevertheless, Li_2O and MgO have not been really considered as possible options due to reactivity and kinetics factors, respectively. On the contrary, CaO is one of the most promising alkaline-earth based materials, with possible real industrial applications. Other interesting materials are ceramics with lithium or sodium phases, which present better thermal stabilities and volume variations than CaO. In addition, the sodium phases may present another advantage if the CO_2 capture is produced in the presence of steam. Under these conditions the sodium phases may produce sodium bicarbonate ($NaHCO_3$) as the carbonated phase, which is twice the amount of CO_2 could be trapped in comparison to the Na_2CO_3 product.

Other ceramics containing alkaline-earth metals are the layered double hydroxides (LDH) or hydrotalcite-like compounds (HTLc). LDHs, also called anionic clays due to their layered structure and structural resemblance to a kind of naturally-occurring clay mineral. These materials are a family of anionic clays that have received much attention in the past decades because of their numerous applications in many different fields, such as antacids, PVC additives, flame retardants and more recently for drug delivery systems and as solid sorbents of gaseous pollutants [22-24]. The LDH structure is based on positively charged brucite like [$Mg(OH)_2$] layers that consist of divalent cations surrounded octahedrally by hydroxide ions. These octahedral units form infinite layers by edge sharing [25]. Due to the fact that certain fraction of the divalent cations can be substituted by trivalent cations at the centers of octahedral sites, an excess of positive charge is promoted. The excess of positive charge in the main layers of LDHs is compensated by the intercalation of anions in the hydrated interlayer space, to form the three-dimensional structure. These materials have relatively weak bonds between the interlayer and the sheet, so they exhibit excellent ability to capture organic or inorganic anions. The materials are easy to synthesize by several methods such as co-precipitation, rehydration-reconstruction, ion exchange, hydrothermal, urea hydrolysis and sol gel, although not always as a pure phase [26].

The LDH materials are represented by the general formula: $[M1-x^{II} Mx^{III}(OH)_2]x+[Am-]x/m \bullet nH_2O$ where M^{II} and M^{III} are divalent (Mg^{2+}, Ni^{2+}, Zn^{2+}, Cu^{2+}, etc.) and trivalent cations (Al^{3+}, Fe^{3+}, Cr^{3+}, etc.), respectively, and A^{m-} is a charge compensating anion such as CO_3^{2-}, SO_4^{2-}, NO_3^-, Cl^-, OH^-, where x is equal to the molar ratio of $[M^{III}/(M^{II} + M^{III})]$. Its value is commonly between 0.2 and 0.33, i.e., the M^{II}/M^{III} molar ratio is in the range of 4 - 2 [25], but this is not a limitation ratio and it depends on the M^{II} and M^{III} composition [27-29].

Among various CO_2 mesoporous adsorbents, LDH-base materials have been identified as suitable materials for CO_2 sorption at moderate temperatures (T ≤ 400 °C) [30-46] due to their properties such as large surface area, high anion exchange capacity (2-3 meq/g) and good thermal stability [37-40]. The LDH materials themselves do not possess any basic sites. For that reason, it is preferred to use their derived mixed oxides, formed by the thermal decomposition of LDH, which do exhibit interesting basic properties. Thermal decomposition of the material occurs in three stages, first at temperatures lower than 200 °C, at which the dehydration of superficial and interlayer water molecules takes place on the material. Then the second decomposition stage takes place in the range of 300-400 °C, at which the structure collapses due to a partial dehydroxylation process, typically associated with both the decomposition of Al-OH and the Mg-OH hydroxides. During dehydroxylation, changes occur in the structure. A portion of the trivalent cations of the brucite like layers migrates to the interlaminar region, allowing the preservation of the laminar characteristics of the material [41]. Finally, the total decomposition of the material occurs at temperatures higher than 400 °C, when the decarbonation process is completed [42].

Once the temperature reaches about 400 °C, LDH forms a three-dimensional network of compact oxygen with a disordered distribution of cations in the interstices, where the cations M^{+3} are tetrahedrally coordinated (interlayer region) and M^{+2} are octahedrally coordinated. The compressive-expansion stresses associated with the formation of the amorphous three-dimensional networks and their connection to the octahedral layer increases the surface area and pore volume, which can help improve the storage capacity properties, for example for gas sorption related applications, besides decreasing the ability of the Mg^{+2} cation to favor physisorption instead of chemisorption [30, 42]. For instance, the thermal evolution of the $Mg/Al\text{-}CO_3$ LDH structure is considered to be crucial in determining the CO_2 adsorption capacity, so there are several studies about this issue [42-44].

Reddy et al. [43] studied the effect of the calcination temperature on the adsorptive capacity of the $Mg/Al\text{-}CO_3$ LDH. They found out that the best properties were obtained at calcination temperature of 400 °C, which they attributed to the obtaining of a combination of surface area and the availability of the active basic sites. Actually, at this temperature the material is still amorphous, which allows having a relatively high surface area. Therefore, there is a high number of exposed basic sites, allowing the reversible CO_2 adsorption according to the following reaction:

$$Mg\text{-}O + CO_2 \rightarrow Mg\text{-}O \dots CO_2 \text{ (ad)} \tag{4}$$

However, if the LDH is calcined under 500 °C, the material is able to transform back to the original LDH structure when it is exposed to a carbonate solution or another anionic containing solution. Finally, if the sample is heated to temperatures above 500 °C, the structural changes become irreversible because of the spinel phase formation [37].

As mentioned, the mixed oxides derived from the LDH calcination possess some interesting characteristics such as high specific surface area, excess of positive charge that needs to be compensated, basic sites and thermal stability at elevated temperatures (200 – 400 °C). Besides these aspects, it is important to consider the advantage of acid-base interactions on the CO_2 sorption applications, where acidic CO_2 molecules interact with the basic sites on the derived oxide. These characteristics make the

LDH-materials acceptable CO_2 captors [43, 45]. However, the CO_2 adsorption capacity of this material is low compared with other ceramic sorbents; reaching mean values smaller than 0.1 mmol/g [46]. Nevertheless, many studies suggest that the adsorption capacity of LDH materials can be improved by modifying a factor set such as: composition, improvement of the material's basicity and contaminant gas stream composition [30-32, 36, 41-45, 47-59].

As previously mentioned, Reddy et al. [43] studied the influence of the calcination temperature of LDHs on their CO_2 capture properties. The Mg_3/Al_1-CO_3 material was calcined at different temperatures ranging from 200 to 600 °C. The results showed that when the calcination temperatures are under 400 °C, LDH is considered to be dehydrated and materials still keep the layered structure intact, wherein the CO_3^{2-} ions are occupying the basic sites. The obtained samples calcined at 400 °C have the maximum BET surface area of 167 m²/g compared with samples calcined at lower temperatures. Moreover, during the calcination of the LDH at higher temperatures (T > 500 °C), most of the CO_3^{2-} decompose to release some basic sites for CO_2 adsorption. However, the final amount of basic sites decreases with the subsequent crystallization of the MgO and spinel ($MgAl_2O_4$). Hence, LDH materials obtained at 400 °C have the highest surface area and the maximum quantities of active basic sites exposed. Because of these characteristics, they achieved a total sorption capacity of 0.5 mmol/g [43]. The same researchers observed that 88% of the captured gas can be desorbed and during the material regeneration 98% of the original weight is gained. This is another important property of LDH materials in high temperature CO_2 separation applications as described later..

As mentioned, the thermal evolution of the layered structure has a great influence on the CO_2 capture. The loss of superficial interlayer water occurs at 200 °C. Then at temperatures between 300 and 400 °C the layer decomposition begins, resulting in an amorphous 3D network with the highest surface area [30], so the adsorption temperature improves the CO_2 capture in the order of 400 > 300 > 20 >200 °C [41-42, 47, 52]

Several researchers have investigated a set of different factors to improve the CO_2 sorption capacity. Yong et al. [47, 48] studied the factors which influence the CO_2 capture in LDH materials, such as aluminum content, water content and heat treatment temperature. Regarding the M/

Al-CO$_3$ LDHs (M = Mg, Ni, Co, Cu or Zn), the best CO$_2$ sorption capacity was obtained for the Mg/Al materials degassed at 400 °C and with adsorption conditions of 25 °C. In general, the sorption capacity follows the trend Ni > Mg > Co > Cu = Zn. However, when the degassed temperature is increased, the trend is modified to Mg (400 °C) > Co (300°C) > Ni (350°C) > Cu (300°C) >Zn (200°C). These results show that Mg/Al-CO$_3$ is the best composition at the degassing temperature of 400 °C [47]. At this temperature, the material consists of an amorphous phase with optimal properties for use as CO$_2$ captor [42]. Also, the influence of Al^{+3} has been studied as a trivalent cation at 25 and 300 °C adsorption temperatures, by Yong [41] and Yamamoto [49] respectively. Both samples were degassed at 300 °C and the results showed that the CO$_2$ capture is influenced by the adsorption temperature. At a temperature of 25 °C, the maximum adsorption was 0.41 mmol/g with an Mg/Al ratio equal to 1.5 (x = 0.375) [41] and at 300 °C the amount of CO$_2$ adsorbed was 1.5 mmol/g for a cation ratio of 1.66 (x = 0.4) [49]. The differences between the two capacities can be attributed to the Al content differences. The Al incorporation in the structure has two functions: 1) to increase the charge density on the brucite-like sheet; and 2) to reduce the interlaminar distance and the number of sites with high resistance to CO$_2$ adsorption [48].

On the other hand, Qian et al. [50] studied the effect of the charge compensation anions (A$^-$ = CO$_3^{-2}$, NO$_3^{-1}$, Cl$^-$, SO$_4^{-2}$ and HCO$_3^{-1}$) on the structural properties and CO$_2$ adsorption capacity of Mg/Al-A$^-$ (molar ratio equal to 3). Despite all of the prepared LDH materials showed the typical XRD patterns of LDH materials, slight structural and microstructural differences were observed. In fact, the interlayer distance changed by varying the interlayer anions due to their difference in sizes and carried charges. These differences affect the morphology and the BET surface area of both fresh and heat-treated LDH materials. Additionally, thermal treatments were performed in order to optimize the adsorption capacity of these materials. The optimal temperature treatment was established for each Mg/Al-A$^-$ based on the surface area of each calcined LDH. Then the CO$_2$ adsorption capacities of calcined LDH were tested at 200 °C. Mg$_3$/Al$_1$–CO$_3$ showed the highest CO$_2$ adsorption capacity (0.53 mmol/g). This value was much higher than those obtained for calcined Mg$_3$/Al$_1$-NO$_3$ > Mg$_3$/Al$_1$-HCO$_3$, Mg$_3$Al$_1$-Cl, and Mg$_3$/Al$_1$-SO$_4$ (≈ 0.1 mmol/g). The results indicated that BET surface area of calcined LDHs seems be the main parameter that determines the CO_2 adsorption capacity because the Mg-O active basic site [43, 45].

It has been demonstrated that the quasi-amorphous phase obtained by the thermal treatment of LDH at the lowest possible temperature has the highest CO_2 capture capacity. This finding is in line with the fact that high calcination temperature can decrease the number of active Mg–O sites due to the formation of crystal MgO [51].

Yong [41] and Yamamoto [49] investigated the influence of the several types of anions. The results suggested that the amounts CO_2 capture decrease as a function of the anion size, which promotes a larger interlayer spacing and the higher charge: $Fe(CN)_6^{4-}$ (1.5 mmol/g) > CO_3^{2-} (0.5 mmol/g) > NO_3^- (0.4 mmol/g) > OH^- (0.4-0.25 mmol/g). The reason is that $Fe(CN)_6^{4-}$ and CO_3^{2-}, because they have more void space in the interlayer due size, and are able to accommodate higher CO_2 quantities. Calcined layered double hydroxide derivatives have shown great potential for high temperature CO_2 separation from flue gases. However, the presence of SO_x and H_2O from flue gases can strongly affect CO_2 adsorption capacity and regeneration of hydrotalcite-like compounds. Flue gases emitted from power stations contain considerable amounts of water in the form of steam. The percentage of water found in the flue gas emitted from different sources varies between 7 and 22%, with the emissions from brown coal combustion having the highest water content [45]. For many other gas adsorption sorbents, steam generally has a negative effect on the adsorption performance because of competition for basic sites between CO_2 and H_2O. However, the presence of water or steam seems to be favorable for the adsorption capacity onto LDH [31,43,53,54]. This fact is the result of the increasing potential energy that is able to further activate basic sites, possibly by maintaining the hydroxyl concentration of the surface material and/or preventing site poisoning through carbonate or coke deposition [31]. An example of the above was reported by Yong et al. [47], who found that water or steam can increase the adsorption capacity of CO_2 by about 25%, from 0.4 mmol/g to 0.5 mmol/g.

Ding et al. [31] studied CO_2 adsorption at higher temperatures (480 °C) under conditions for steam reforming of methane. They found an adsorption capacity of 0.58 mmol/g, which was independent of water vapor content in the feed. In turn, Reddy et al. [45] investigated calcined LDHs' sorption performance influenced by CO_2 wet-gas streams. LDH samples were calcined at 400 °C [43] before measuring CO_2 sorption at 200 °C. The gas streams used were CO_2, CO_2 + H_2O, flue gas (14% CO_2, 4% O_2 and 82 % N_2) +12% H_2O.

For a pure CO_2 dry sorption, the maximum weight gain was 2.72% (~0.61 mmol/g) after 60 min, whereas the wet adsorption increased the weight of the calcined LDHs to 4.81%, showing an additional 2.09%, where He and He + H_2O were used to remove the H_2O water capture. The results showed that the helium has virtually no significant sorption affinity for the material, whereas the water-sorption profile of it clearly indicates a water weight gain of 1.67%, i.e., the gain was 0.1mmol/g due to steam presence, showing that water has a positive effect, shifting the CO_2 sorption by 0.42% as compared to dry CO_2 sorption. Also, these results revealed that in all cases about 70% sorption occurs during the first 5 min and reaches equilibrium after around 30 min.

To determine the influence of CO_2, Reddy et al. [43] tested a sample in both, wet and dry CO_2 stream conditions. The experiments showed that the same quantity of CO_2 can be trapped for the solid sorbent after two hours. The presence of water in the stream only affects the kinetics of the process. This result is in agreement with that reported by Ding et al. [31]. On the other hand, the results of the material tested suggest that the fact the CO_2 capture from flue gas was higher than in a pure stream of CO_2 might have been because the polluted gas was diluted in the stream. The presence of the water does not enhance de CO_2 capture; the maximum CO_2 adsorbed was 0.9 mmol/g. The differences between Reddy et al. results and the previously mentioned studies can be caused by the use of uncalcined LDHs, which already contain an -OH network.

To apply these materials in industrial processes, it is important to know the times during which each sorbent material can be used. Tests of the cyclability in LDH materials disclose that as function of the temperature the CO_2 capture time can vary. This can be attributed to CO_2 chemisorbed during each cycle [54] and/or to the formation of spinel-based aluminas, such as γ-Al_2O_3 (at temperatures higher than 400 °C). Hibino et al. [52] found that the carbonate content, acting as charge-compensating anion, continuously decreases in subsequent calcination–rehydration cycles. Reddy et al.

tested LDH materials during six CO_2 adsorption (200 °C)-desorption (300 °C) cycles. The average amount gained was 0.58 mmol/g, whereas 75% of this value is desorbed, reaching desorption equilibrium after the third cycle. This can be attributed to the stabilization of the material phase and basic sites during the temperature swing.

Hufton et al. [54] studied a LDH material during several cycles in dry and wet CO_2 flows. As previously discussed, the presence of steam in the flow gas improves the CO_2 adsorption. However, after 10 adsorption cycles, the capture decreased 45%. The same behavior was observed in the dry gas flow. However, the final capture was similar to the wet gas stream, in agreement with Reddy et al. [43].

Recent studies have demonstrated that K-impregnated LDH or K-impregnated mixed oxides have a better CO_2 capture capacity due to the addition of K alkaline-earth element that improves the chemical affinity between the acidic CO_2 and alkaline surface of the sorbent material [32, 36, 55-56]. Additionally, it has been proposed that K-impregnation reduces the CO_2 diffusion resistance in the material. [57]. Hufton et al. [58] showed that the K-impregnation increases the CO_2 capture, but there is an optimal quantity of K to reach the maximum capture. Qiang et al. [50] tested an Mg_3/Al_1-CO_3 (pH = 10) impregnated with 20 wt.% K_2CO_3. The CO_2 adsorption capacity was increased between 0.81 and 0.85 mmol/g in the temperature range of 300–350 °C. This adsorption capacity is adequate for application in water gas shift reactions (WGS).

Lee et al. [59] tested the behavior of three commercial LDHs impregnated with K (K_2CO_3/LDH ratio between 0 and 1). Three Mg/Al-CO_3 LDH with different contents of magnesium were used. Results indicated that the sorption capacity of the LDH is improved by about 10 times with the optimal K_2CO_3 additions. Additionally, it was observed that impregnation is not the only factor that influences the adsorption but the composition too. The best value was obtained when the content of divalent cation was reduced and therefore, the material had a composition with the maximum trivalent cation content. The CO_2 adsorption capacity improved from 0.1mmol/g to 0.95mmol/g with K_2CO_3/LDH weight ratio equal to 0.35 at 400 °C. After determining the optimal alkaline source/LDH ratio, a set of samples was evaluated as a function of the temperature and the results showed a maximum of 1.35 mmol/g, at 50 °C. In the impregnated materials, CO_2 chemisorption can occur and the sorbed CO_2 can be further stored as metal carbonate forms.

Other alkaline elements can be used to improve the sorption capacity of materials. Martunus et al. [46] studied the impregnation of LDH with Na and K. The LDH samples were thermally treated at 450 °C for 5 min then calcined samples were re-crystallized in K_2CO_3 and Na_2CO_3 (1 M) solutions. The re-crystallized materials were tested as CO_2 captors and the capture was maximum with LDH-Na (0.688 mmol/g) > LDH-K (0.575 mmol/g) at 350 °C after five cycles. Finally, the re-crystallized material with the highest capture was calcined at 650 °C for 4 h and re-crystallized with a solution containing the appropriate quantities of K and Na to achieve alkaline metal loading up to 20%. When the sample was Impregnated with additional K and Na at 18.4% and 1.6%, respectively, the adsorption capacity rose from 0.688 to 1.21 mmol/g. This capacity increase was achievable despite the relatively low BET surface area, equal to 124 m^2/g.

Other alkaline elements such as cesium have been studied as reinforcement. Oliveira et al. [55] tested commercial Mg_1/Al_1-CO_3 and Mg_6Al_1-CO_3 impregnated with K and Cs carbonates. The materials were evaluated in the presence of steam (26% v/v water content) gas at different temperatures (306, 403 and 510 °C) at 0.4 bar of CO_2 partial pressure (total pressure 2 bar). The LDH with the highest sorption capacity was Mg_1/Al_1-CO_3–K with 0.76 mmol/g at 403 °C. Among the Cs impregnated samples, the Mg_6Al_1-CO_3-Cs presented the highest capacity with 0.41 mmol/g, while the commercial LDH samples presented CO_2 sorption capacities around 0.1 mmol/g.

The results suggest the existence of a sorption mechanism combining physical adsorption and chemical reaction. First the maximum physical adsorption is reached, then the chemisorption begins, but there is an optimal temperature. If the temperature is too low, the chemisorption does not happen, but with higher temperatures the loss of porosity impedes the contact of CO_2 molecules with active basic sites promoted by the alkaline element addition.

These results suggest there is an optimum amount of K_2CO_3 to impregnate the LDH that achieves a balance between the increase in the basicity of the sorbent material and its reduction of surface area, associated with

CO_2 capture capacity. The influence of potassium is currently not clear and the relevant research is still ongoing. Finally, CO_2 adsorption capacity on the synthesized 20 wt.% $K_2CO_3/Mg_3/Al_1-CO_3$ (pH = 10) probably could be further increased in the presence of steam.

10.3 CERAMIC OXIDE MEMBRANES AS AN ALTERNATIVE FOR CO_2 SEPARATION

Membrane-based processes, related to gas separation and purification, have achieved an important level of development for a variety of industrial applications [60]. Therefore, the use of separation membranes is one of the promising technologies for reducing the emissions of greenhouse gases such as CO_2. The term membrane is defined as a permselective barrier between two phases, the feed or upstream and permeate or downstream side [61]. This permselective barrier has the property to control the rate of transport of different species from the upstream to the downstream side, causing the concentration or purification of one of the species present in the feed gas mixture.

Membrane-based processes offer the advantage of large scale application to separate CO_2 from a gas mixture. Figure 2 schematizes the process where concentrated CO_2 is selectively separated from flue gas that is mainly composed of nitrogen and carbon dioxide along with other gases such as water vapor, SO_x, NO_x and methane. Subsequent to the membrane process, concentrated CO_2 obtained at the permeate side can be disposed or used as raw material for the synthesis of several chemicals such as fuel and value-added products [62].

Of course, the rate of transport or permeation properties of a particular gas through a given membrane depend on the nature of the permeant gas, as well as the physical and chemical properties of the membrane.

Inorganic membranes are more thermally and chemically stable and have better mechanical properties than organic polymer membranes; ceramic membranes offer both the advantage of large scale application and potential for pre- and post-combustion CO_2 separation applications, where membranes systems would be operating at elevated temperatures of 300-1000 °C [63].

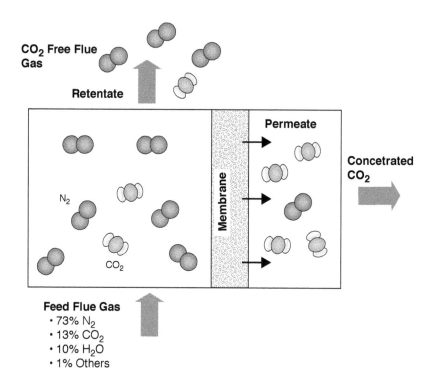

FIGURE 2: Membrane-based processes for the carbon dioxide separation from flue gases. The concentrated CO_2 is obtained in the permeate side.

Inorganic ceramic membranes can be classified as porous and nonporous or dense. These differ from each other not only in their structures but also in the mechanism of permeation. In porous membranes, the transport of species is explained with the pore-flow model, in which permeants are transported by pressure-driven convective flow through the pore network. Separation occurs because one of the permeants is excluded (molecular filtration or sieving) from the pores in the membrane and remains in the retentate while the other permeants move towards the downstream side. On the other hand, in nonporous membranes, separation occurs by solution-diffusion, in which permeants dissolve in the membrane material and then diffuse through the bulk membrane by a concentration gradient [60].

10.3.1 POROUS MEMBRANES BASED ON ALKALINE AND ALKALINE-EARTH CERAMIC OXIDES FOR CO_2 SEPARATION

Among the porous systems for CO_2 separation, both microporous (carbon, silica and zeolite membranes) and modified mesoporous membranes have been reported [63-64].

Zeolites are hydrous crystalline aluminosilicates that exhibit an intracrystalline microporous structure as a result of the particular three-dimensional arrangement of their TO_4 tetrahedral units (T = Si or Al) [65]. Zeolite membranes are commonly prepared as thin films grown on porous alumina supports via hydrothermal synthesis and dry gel conversion methods [66]. Zeolite membranes of different structures have been developed to separate CO_2 from other gases via molecular sieving [67-69]. For example, membranes prepared with the 12-member ring faujasite (FAU)-type zeolite show high separation factors of 20-100 for binary gas mixtures of CO_2/N_2 [69]. In the same sense, T zeolite membranes exhibited very high selectivity, of about 400, for CO_2/CH_4 and 104 for CO_2/N_2. The high selectivity of CO_2/CH_4 exhibited by T zeolites is due to the small pore size of about 0.41 nm, which is similar in size to the CH_4 molecule but larger than CO_2 [69]. Table 1 shows the kinetic diameter of various molecules that are present in CO_2 containing gas mixtures such as flue and natural gas [70].

TABLE 1: Kinetic diameter of various molecules based on the Lennard Jhones relationship.

Molecule	Kinetic diameter ($\mathring{A}$)
H_2O	2.65
H_2	2.69
CO_2	3.3
O_2	3.46
N_2	3.64
CH_4	3.80

Deca-dodecasil 3R (DDR) (0.36 nm x 0.44 nm), and pseudo-zeolite materials like silicoaluminophosphate (SAPO)-34 (0.38 nm) also show high CO_2/CH_4 selectivities due to narrow molecular sieving, which controls molecular transport into this material [69, 71-73]. For example, Tomita et al. [74] obtained a CO_2/CH_4 separation factor of 220 and CO_2 permeance values of 7 x 10^{-8} mol m^{-2} s^{-1} Pa^{-1} at 28 °C on a DDR membrane [75].

As discussed, one of the most important factors controlling permeation through microporous membranes is the restriction imposed by the molecular size of the permeant. However, the transport mechanism in microporous systems is more complex than just size exclusion and the permeation and selectivity properties are also affected by competitive adsorption among perment species that produce differences in mobility [76].

Thus, the diffusion mechanism for gas permeation through microporous membranes can be characterized by two modes: one controlled by adsorption and a second one where diffusion dominates [63]. In the case of adsorption-controlled mode with permeating gases having strong affinity with the membrane, a gas permeation flux equation is obtained by assuming steady-state single gas permeation, a constant diffusivity and a single gas adsorption described by a Langmuir-type adsorption isotherm, as in Eq. (5).

$$J = \phi q s D c L 1 + b P f 1 + b P p \text{ or } J = \phi q s D c L 1 - \theta p 1 + \theta f \tag{5}$$

where J is the permeation flux, ϕ is a geometric correction factor that involves both membrane porosity and tortuosity, Dc is the corrected diffusivity of the permeating species, L is the membrane thickness, Pf and Pp represent the feed and permeate pressure respectively and θf and θp represent the relative occupancies.

Furthermore, if the adsorption isotherm of the permeating gas is linear (1 >> bP), then flux permeation is described by Eq. (6).

$$F = \phi q s D c L D c L K \tag{6}$$

where F is the permeance and K = qsb is the adsorption equilibrium constant. Therefore, from Eq. (5) it can be concluded that permeance is determined by both diffusivity (Dc) and adsorption (K). Based on the above, an interesting option to enhance membrane properties is to intercalate zeolite membranes with alkaline and alkaline-earth cations. Zeolite intercalation can enhance the separation between CO_2 and other molecules such as N_2 by promoting preferential CO_2 adsorption [63, 77]. It is well known that zeolites show affinity for polar molecules, like CO_2, due to the strong interaction of their quadrupole moment with the electric field of the zeolite framework. In this sense, the adsorption properties of zeolites can be enhanced by the inclusion of exchangeable cations within the cavities of zeolites where the adsorbent-adsorbate interactions are influenced by the basicity and electric field of the adsorbent cavities [78-80]. Lara-Medina et al. [77] carried out separation studies of CO_2 and N_2 with a silicalite-1 zeolite membrane prepared via hydrothermal synthesis and subsequently modified by using lithium solutions in order to promote preferential CO_2 adsorption and diffusion. CO_2/N_2 separation factor increases from 1.46 up to 6 at 25 psi and 400 °C after lithium modification. An et al. [79] studied a series of membranes prepared starting from natural Clinoptilolite zeolite rocks. Disk membranes were obtained by cutting and polishing of the original minerals, which were subsequently chemically treated with aqueous solutions containing Li, Na, Sr or Ba ions. Ionic exchanged membranes showed better permeation properties due to the presence of the extra framework cations.

Although zeolite membranes offer certain advantages in comparison with polymer membranes, such as chemical stability, the main issues are related to the selectivity decrease as a function of the permeation temperature. This is explained in terms of the contribution of the adsorption to the separation, which decreases sharply as temperature increases. At high temperature, physical adsorption becomes negligible and permeation is mainly controlled by diffusion [63, 76]. Additionally, due to the fact that CO_2 and N_2 molecules have similar sizes (Table 1), the difference in diffusivity is not a strong controlling factor in determining selectivity.

Modified γ-Al_2O_3 mesoporous membranes have been also reported as a means for CO_2 separation [64]. Transport mechanisms in porous membranes have the contribution of different regimes. An overview of the different mechanisms is given in Table 2.

TABLE 2: Transport mechanisms in porous membranes.

Transport Type	Pore diameter	Characteristics
Viscous flow	>20 μm	Non selective.
Molecular diffusion	>10 μm	Affects the total flow resistance of the membrane system.
Knudsen diffusion	2 – 100 nm	Occurs when the mean free path of the molecule is much larger than pore radius of the membrane.
		Shows selectivity based on molecular weights.
Surface diffusion		
Capillary condensation		Shows selectivity due to interaction of molecules with membrane walls.
Micropore diffusion (Configurational diffusion)	< 1.5 nm	

Depending on the particular system, permeability of a membrane can involve several transport mechanisms that take place simultaneously. Considering no membrane defects and pore sizes in the range of 2.5-5 nm, γ-Al$_2$O$_3$ based membranes theoretically have two transport regimes: Knudsen diffusion and surface diffusion. Eq. (7) describes the permeability of a membrane by taking into consideration the Knudsen and surface diffusion.

$$F = 2\varepsilon\mu r3RTL8RT\pi M0.5 + 2\varepsilon\mu DsrA0NavdxsdP \tag{7}$$

where r is the mean pore radius, μ is a shape factor, R is the universal gas constant, T is the temperature, P is the mean pressure, M is molar mass of the gas, Ao is the surface area occupied by a molecule, Ds is the surface diffusion coefficient, Nav is Avogadro's constant and Xs is the percentage of occupied surface compared with a monolayer [81].

For the cases when Knudsen diffusion dominates, selectivity can be correlated to the molecular weights of the permeating gases by the so called Graham's law of diffusion, which establishes that the transport rate of any gas is inversely proportional to the square root of its molecular weight. The CO_2/N_2 separation factor considering pure Knudsen diffusion is given by Eq. (8) and has a value of just 0.8. Therefore, Eq. (8) clearly shows that separation via Knudsen is limited for systems where species are of similar molecular weight.

$$\alpha CO_2 N_2 = MCO_2 MN_2 \tag{8}$$

Based on the aforesaid, CO_2/N_2 separation factor can be better enhanced by promoting the surface diffusion mechanism (second term on the right hand side of Eq. (7)). Surface diffusion involves the adsorption of gas molecules on the surface of the pore and subsequent diffusion of the adsorbed species along the surface by a concentration gradient. Then separation properties of a membrane can be improved by generating such an interaction between one component of the feed gas mixture with the membrane; one option being via a chemical modification.

Cho et al [81] prepared a series of thin (2-5 μm thickness) γ-Al_2O_3 and CaO- or SiO_2-modified γ-Al_2O_3 membranes for CO_2 separation at temperatures between 25 and 400 °C. Impregnation of membranes with SiO_2 or alkaline CaO was done in order to improve the CO_2/N_2 selectivity by promoting adsorption between CO_2 gas molecules and the membrane pore wall. Although this kind of chemical modification of the membrane surface and the pore walls is able to activate the surface diffusion mechanism, an interesting behavior was observed. The CO_2/N_2 separation factor increased from 1.0 to 1.38 at 25 °C after modification of the γ-Al_2O_3 with SiO_2. On the other hand, CaO impregnated membranes showed a separation factor of 0.98, which is even lower than that of the unmodified γ-Al_2O_3. The same behavior has been reported by Uhlhorn et al. [82-83]. They reported MgO modified γ-Al_2O_3 membranes which did not show significant enhancement in the permeation and selectivity properties as a result of the modification process. This fact was explained in terms of the surface diffusion mechanisms. As discussed, it is expected that physico-chemical modifications of the membrane can enhance preferential adsorption of the gas species in the feed. Impregnations with alkaline oxide such as calcium oxide or magnesia on the γ-alumina surface give more strong base sites than those promoted by silica. Therefore, it promotes a strong bonding of CO_2 on the alumina surface, causing CO_2 molecules to lose mobility, resulting in a smaller contribution of surface diffusion to the total transport mechanism.

There is another kind of membrane where alkaline and alkaline-earth ceramic oxides have been used for the fabrication of CO_2 permselective membranes. In these cases ceramic materials were chosen because of their well-known properties of physisorption of CO_2 at low and intermediated temperatures.

Kusakabe et al. [84] prepared both pure and modified $BaTiO_3$ CO_2 permselective membranes via the alkoxide based sol-gel method; impregnation and calcination at 600 °C. In order to establish the effects of CO_2 partial pressure, temperature and influence of the secondary oxide presence (CuO, MgO or La_2O) on the CO_2 adsorption properties of the membranes, pure and modified barium titanate powders were first evaluated by thermogravimetry and chromatography techniques. Dynamic CO_2 absorption was evaluated by applying the impulse response method, wherein the $BaTiO_3$ powder was packed in a separation column. The results suggested that the CO_2 molecules adsorbed on the $BaTiO_3$ powder are mobile at temperatures about 500 °C. Therefore, this membrane exhibits CO_2 permeation due to surface diffusion mechanism. Even though the prepared membranes showed selectivity, the Knudsen diffusion still has an important contribution to the gas transport due to the presence of membrane defects. The maximum separation factor of CO_2/N_2 through the membranes was estimated as 1.2. Therefore, further improvement of the permeation properties of this kind of membrane requires obtaining pinhole-free membranes.

Based on the same criteria, Nomura et al. [85] prepared Li_4SiO_4-based thin membranes on porous alumina supports. Membranes were obtained by the thermal treatment of different silica containing porous materials (Silicalite-1 and mesoporous silica) impregnated with lithium compounds. The authors called this method solid conversion. The use of different silica porous sources was proposed in order to enhance the reaction rate of Si and Li on the porous support at relatively low temperature, avoiding the reaction between the Li and alumina support itself. In the case of Silicalite-1 (MFI zeolite), a zeolite thin film was first prepared on the support by following the dry gel conversion technique. Then, the prepared Silicalite-1 layer was impregnated via dipping into a slurry containing lithium and silica fumed reactants (Li:Si = 4:1) and subsequently into a Li_2CO_3-K_2CO_3 slurry. The membrane was finally calcined at 600 °C for 2 h. It is believed that carbonate melts to fill the cracks and the pinholes of the Li_4SiO_4 formed membrane. A similar procedure of coating and calcination was carried out to prepare high quality membranes starting from mesoporous silica sources with pore sizes of 1.8-12.8 nm. Precursors react to form a Li_4SiO_4 membrane of 2-5 μm thickness that exhibits an N_2 permeance of 1.8 x 10^{-9} mol m^{-2} s^{-1} Pa^{-1} at 400 °C. This suggests there are no big defects after impregnation of the membrane with the binary mixture of Li_2CO_3-K2CO3carbonate. Due to the fact that the membrane operates in a rich CO_2 atmosphere, carbonates do not decompose even at temperatures of 600 °C. The maximum CO_2/N_2 permeance ratio was 0.85. The separation factor was higher than that for the Knudsen diffusion. Therefore, it can be conclude that Li_4SiO_4 layer was selective to CO_2 over N_2 at high temperature of 600 °C.

Nomura [86] reported a two-stage approach for the preparation of Li_4SiO_4-CO_2 selective membranes that involves the fabrication of a supported Li_4SiO_4 membrane and its subsequent modification by using a chemical vapor deposition (CVD) method. First, for the preparation of a thin Li_4SiO_4 membrane the so called solid conversion method described before was used, which is based on the reaction between a porous silica source and a lithium containing solution coated on a porous alumina membrane support. Although the formed membranes showed certain selectivity due to the preferential adsorption of CO_2 over N_2, the presence of pinholes and cracks caused low separation factors. Therefore, the membrane defects were fixed by using the counter diffusion CVD method to form a silica coating that fills the gaps between the lithium orthosilicate particles that make up the membrane. N_2 permeance was reduced about three orders of magnitude after CVD modification. Nitrogen permeance before and after the CVD treatment was 3.4 x 10^{-6} mol m^{-2} s^{-1} Pa^{-1} and 1.2 x 10^{-9} mol m^{-2} s^{-1} Pa^{-1} respectively. In the same sense, the CO_2/N_2 permeance rate increased from 0.7 to 1.2 at 600 °C. Some issues related with this system are the chemical and structural stability of the membranes observed during the permeation tests at elevated temperature. The membranes were broken when permeation tests were carried out at temperatures higher than 700 °C, with the consequent decrease in the CO_2/N_2 selectivity. The aforesaid is the result of the CO_2 chemisorption on the membrane. Lithium ortho-silicate reacts with CO_2 to form lithium carbonate and lithium metasilicate (Li_2SiO_3) as products, as indicated by Eq. (9).

$$Li_4SiO_4 + CO_2 \leftrightarrow Li_2CO_3 + Li_2SiO_3 \qquad\qquad (9)$$

Thermodynamically, this reaction is prone to occur at temperatures between room temperature and about 700 °C. However, experimentally it has been observed that reaction kinetics sharply increase above 550 °C. At these temperatures, the formation of carbonates involves an important change in volume that ends in the membrane's rupture.

Therefore, one of the issues related to the development of this kind of inorganic membrane is the thermochemical stability. Due to reactivity of alkaline and alkaline-earth ceramic oxides with CO_2 to form carbonates, not only preferential adsorption of CO_2 molecules over N_2 occurs, but CO_2 chemisorption and reaction.. Therefore, it is mandatory to establish the operational temperature within a range where CO_2 selective adsorption on the membrane layer promotes the separation process without reaction.

10.3.2 NONPOROUS MEMBRANES BASED ON ALKALINE AND ALKALINE-EARTH CERAMIC OXIDES FOR CO_2 SEPARATION

Some researchers have proposed the use of alkaline and alkaline-earth ceramic oxides to prepare membranes that are able to separate CO_2 at high temperatures via a different transport mechanism than those observed on porous membranes. Li_2ZrO_3 and Li_4SiO_4 based membranes are examples of the aforesaid. Permselectivity of CO_2 through these membranes takes place not only due to the selective CO_2 adsorption properties of ceramic phases but also via a mechanism of gas separation that involves the transport of CO_3^{2-} and O^{2-} ionic species through the electrolytes (carbonate-metal oxide) phases formed by the reaction of the membrane with the CO_2 [87-89].

Kawamura et al. [87] fabricated and characterized a membrane for CO_2 separation at high temperatures. The membrane was made of lithium zirconate (Li_2ZrO_3), an alkaline ceramic oxide that reacts with CO_2 to produced Li_2CO_3 and ZrO_2. These two reaction products are electrolyte materials produced in-situ when the membrane is exposed to the rich carbon dioxide atmosphere. The electrolytes formed thus are capable to transport both CO_2 and O_2 across the membrane via a dual ion conduction mechanism. The prepared membrane exhibited a separation factor of 4.9 between CO_2 and CH_4 gas molecules at a temperature of 600 °C. The obtained separation factor is higher than the Knudsen diffusion limit, 0.6. Therefore, the results clearly suggest the potential use of this kind of membrane system for CO_2 separation such as the case of CO_2 removal from natural gas.

Yamaguchi et al. [88] investigated the concept of the dual-ion conduction facilitated mechanism previously observed for the case of Li_2ZrO_3 membranes by focusing their efforts on the preparation of a CO_2 permselective membrane based on lithium orthosilicate (Li_4SiO_4). The supported membrane was prepared via a dip coating technique by using Li_4SiO_4 suspensions. The coating process was repeated several times before impregnation of the membrane with a Li_2CO_3/K_2CO_3 carbonate mixture and final sintering at 750 °C. In this membrane system, Li_4SiO_4 reacts in-situ with CO_2 to form Li_2CO_3 and Li_2SiO_3.

Gas separation studies were performed by using CO_2/N_2 mixtures as feed gas. The observed CO_2 permeance values were about 1×10^{-8} mol $m^{-2}s^{-1}Pa^{-1}$ in the temperature range of 525-625 °C. The CO_2/N_2 separation factor was estimated between four and six. Figure 3 shows a scheme of the dual-ion conduction mechanism explained as follows. In the feed side, carbon dioxide dissolves in the material and diffuses as carbonate ions through the molten carbonate electrolyte due to a concentration gradient. Then, in the downstream side of the membrane, the formation of gaseous CO_2 implies the formation of oxygen ions which must diffuse back to the feed side across the membrane and apparently through the formed Li_2SiO_3 skeleton to obtain the charge balance.

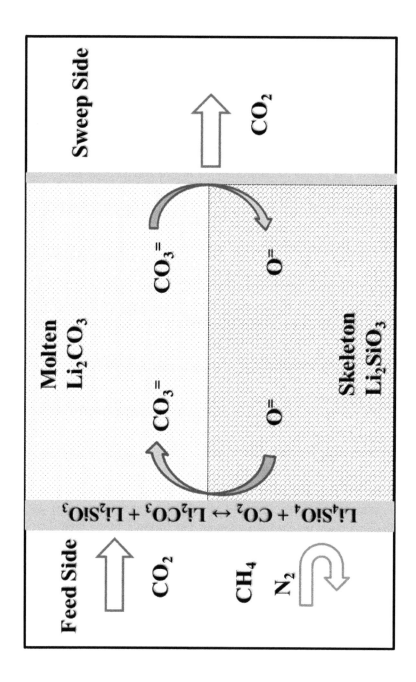

FIGURE 3: Schematic representation of a membrane system for the CO_2 separation via a dual-ion conduction mechanism.

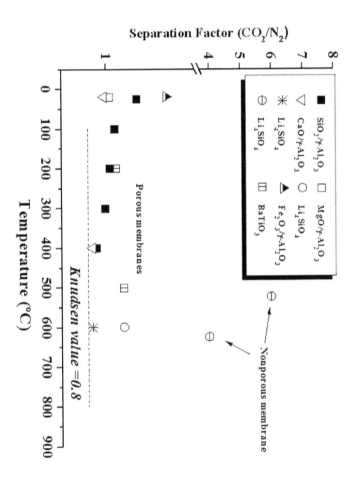

FIGURE 4: CO$_2$/N$_2$ separation factor of different ceramic oxide membranes.

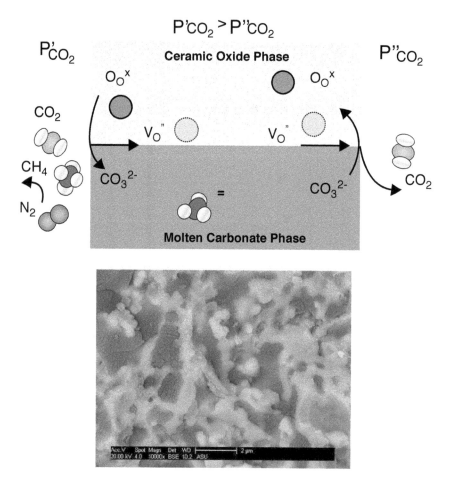

FIGURE 5: Schematic representation of a membrane system for the CO_2 separation and SEM image of a ceramic oxide-carbonate dual-phase membrane

The proposed transport mechanisms supports the higher selectivity values observed in the permeation test for both systems, Li_2ZrO_3 and Li_4SiO_4. Figure 4 shows the separation factor values (CO_2/N_2) obtained for different ceramic membranes described in the present report. The pure Knudsen value is written as baseline and separation factor of nonporous Li_4SiO_4 for comparison purposes. However, it is important to mention that the oxygen ion diffusion process is not totally clear. Indeed, there is no experimental study regarding the oxygen ionic conductivity properties of Li_2SiO_3 phase. On the other hand, pure ZrO_2 exhibits poor bulk oxygen ion conductivity. In fact, good conduction properties are observed only in acceptor-doped ZrO_2 based materials with oxygen vacancies being the predominant charge carriers [90]. Therefore, oxygen ion conduction through the membrane must be related to different transport paths, such as grain boundaries and interfacial regions formed between the ceramic and molten carbonate on the membrane.

More recently, the promising concept of ceramic oxide-carbonate dual-phase membranes has been proposed for carbon dioxide selective separation at intermediate and high temperatures (450-900 °C) [91-97].

This concept involves the fabrication of nonporous membranes capable of selectively separating CO_2 via its transport, as carbonate ions. Dual phase membranes are made of an oxygen ion conductive porous ceramic phase that hosts a molten carbonate phase. Rui et al. [98] proposed the CO_2 separation by the electrochemical conversion of CO_2 molecules to carbonate ions (CO_3^{2-}), which are subsequently transported across the membrane. Carbonate ionic species (CO_3^{2-}) are formed by the surface reaction between CO_2 and oxygen that comes from the ceramic oxide phase (feed side, Eq.(10)) and then transport of CO_3^{2-} takes place through the molten carbonate.

$$CO_2 + OOx \leftrightarrow CO_3^{2-} + VO^{\cdot\cdot} \tag{10}$$

Once carbonate ions have reached the permeate side, molecular CO_2 is released to the gas phase, delivering OOx species back to the ceramic oxide solid phase. This process takes place due to a chemical gradient of CO_2 in the system (Figure 5). Here, it is important to emphasize that dual-phase membranes are nonporous and therefore exhibit high separation selectivity as a result of the transport mechanism. Figure 5 also shows the SEM image of the cross section of a ceramic oxide-carbonate membrane prepared by pressing $La_{0.6}Sr_{0.4}Co_{0.8}Fe_{0.2}O_{3-\delta}$ powders and subsequent infiltration of the obtained porous ceramic (bright phase) with carbonate (dark phase).

Table 3 summarizes the different studies reported and certain advances that have been achieved so far regarding the dual-phase membrane concept. This table also includes the Li2ZrO3 and Li4SiO4 nonporous membranes previously described. Although the original reports do not clearly explain the operational mechanism [26-27], the dual-phase membrane concept gives a much better idea of the possible phenomenology involved [30,33,36].

TABLE 3: Reported studies on dual-phase and related membranes for CO_2 separation.

Ceramic Oxide phase	Molten Carbonate phase	Membrane features	Preparation method	Permeance (mol s⁻¹m⁻²Pa⁻¹)	Separation Factor (CO_2/N_2)	Ref.
Li_2ZrO_3	Li_2CO_3	Thick membrane	In situ by exposing Li_2ZrO_3 to CO_2 atmosphere	1×10^{-8}	4.9 (CO_2/CH_4 at 600°C)	[87]
Li_4SiO_4/Li_2SiO_3	K_2-Li_2CO_3	Thin supported membrane	Impregnation of carbonate	2×10^{-8}	5.5 (at 525°C)	[88]
$La_{0.6}Sr_{0.4}Co_{0.8}Fe_{0.2}O_{3-\delta}$	Li-Na-K_2CO_3	Thick membrane (0.35-1.5 mm)	Pressing and direct infiltration	4.77×10^{-8}	225 (at 900°C)	[91]
8 mol% Yttria doped zirconia (YSZ)	Li-Na-K_2CO_3	Thin freestanding membranes (200-400 μm)	Tape casting and in situ infiltration	2.0×10^{-8} (YSZ)	> 2 (at 800 °C)	[92]
10 mol% Gadolinia doped ceria (GDC)	Li-Na_2CO_3			3.0×10^{-8} (GDC)		
$Ce_{0.8}Sm_{0.2}O_{1.9}$	Na_2-Li_2CO_3	Thick membrane (1.2 mm)	Pressing of SDC-NiO powders where NiO is a sacrificial template	$\sim 1.2 \times 10^{-6}$	155-255 (at 700°C)	[93]

TABLE 3: *Cont.*

Ceramic Oxide phase	Molten Carbonate phase	Membrane features	Preparation method	Permeance (mol s^{-1}m^{-2}Pa^{-1})	Separation Factor (CO_2/ N_2)	Ref.
$Bi_{1.5}Y_{0.3}Sm_{0.2}O_3$	Li-Na-K_2CO_3	Thin supported membrane (50 μm)	Dip coating of modified thick support and infiltration	1.1 x 10^{-8}	2(at 650°C)	[94]
8 mol% Yttria doped zirconia (YZS)	Li-Na-K_2CO_3	Thin supported membrane (10 μm)	Dip coating of YSZ on nonwettable thick support and infiltration	~ 7.8×10−8	---	[95]
$Ce_{0.8}Sm_{0.2}O_{1.9}$	Li-Na-K_2CO_3	Thin tubular membrane	Centrifugal casting and direct infiltration			[96]
$La_{0.6}Sr_{0.4}Co_{0.8}Fe_{0.2}O_{3-\delta}$	Li-Na-K_2CO_3	Thick disk-shaped membrane	Pressing and direct infiltration			[97]

10.3.3 APPLICATIONS OF CO_2 PERMSELECTIVE CERAMIC OXIDE MEMBRANES FOR THE DESIGN OF MEMBRANE REACTORS.

As mentioned, CO_2 can be used as raw material for the synthesis of several chemicals [99]. Moreover, if CO_2 is concentrated or separated by a membrane system exhibiting high CO_2 permeation and permselectivity, this open up the possibility to develop a continuous process of membrane reaction to simultaneously capture and chemically convert CO_2. For example, if the membrane is able to separate CO_2 at intermediate and even high temperatures, it can be used for the design of a membrane reactor for the production and purification of hydrogen and syngas. Syngas is a gaseous fuel with a main chemical composition of CO, H_2, CO_2, and CH_4. Syngas can be used as feedstock for the synthesis of several other clean fuels such as H_2, methanol, ethanol, diesel and other hydrocarbons synthesized via the Fischer-Tropsch process [100-104].

Among the different processes for the synthesis of syngas and hydrogen, CO_2 methane reforming Eq. (11) and the water-gas shift reaction (WGS) Eq. (12) are the most promising options.

$$CH_4 + CO_2 \leftrightarrow 2CO + 2H_2 \tag{11}$$

$$CO + H_2O \leftrightarrow CO_2 + H_2 \tag{12}$$

Figure 6 schematizes the membrane reactor concept considering the two reactions described above. Figure 6A shows a membrane reactor for dry reforming of methane to produce syngas at temperatures between 700 and 800 °C. Figure 6B illustrates the use of ceramic oxide membranes for hydrogen purification by separating the CO_2 from water-gas shift products at about 550 °C. Additionally, Figure 6B shows the possibility of using a ceramic sorbent to chemically trap the permeate CO_2 and therefore enhance the CO_2 permeation process by reducing the concentration of CO_2 in the permeate side.

10.4 CHEMICAL TRANSFORMATION OF CO2 CATALYZED BY CERAMIC MATERIALS: THE USE OF NEW ALTERNATIVES.

One of the most widely used chemical absorption techniques for carbon capture and storage/sequestration (CCS) is CO_2 adsorption by ceramic materials. Once CO_2 has been captured-fixed, it can be converted into value-added products such as precursors in chemical transformation reactions. CO_2 is extensively used for enhanced oil recovery, as a monomer feedstock for urea and polymer synthesis, in the food and beverage industry as a propellant, and in production of chemicals. Therefore, the capture-fixation of CO_2 would make a system suitable for accomplishing chemical transformation of CO_2. The utilization of carbon dioxide is also very attractive because it is environmentally benign [105-115]. CO_2 conversion to fuel and value-added products is an ideal route for CO_2 utilization due to the simultaneous disposal of CO_2 and the benefit that many products can be used as alternate transportation fuels [116]. CO_2 chemical transformation methods include (i) reverse water-gas shift, (ii) hydrogenation to hydrocarbons, alcohols, dimethyl ether and formic acid, (iii) reaction with hydrocarbons to syngas, (iv) photo- and electrochemical/catalytic conversion, and (v) thermo-chemical conversion [100-122].

CO_2 can be catalyzed to valuable organic or inorganic compounds, where some basic catalytic materials (containing alkaline or alkaline-earth elements) are used. The activation of CO_2 by alkali metals has received considerable attention in various surface science studies, which have demonstrated the formation of intermediate CO_2, dissociation of CO_2 and formation of oxalate and carbonate alkali compounds [118-121]. Carbon dioxide has been identified as one such potential vector molecule (through reduction to syngas, methanol, methane, formic acid, formaldehyde, dimethylether (DME) and short-chain olefins) [117-118, 120-122]. CO_2 is a kinetically and thermodynamically stable molecule, so CO_2 conversion reactions are endothermic and need efficient catalysts to obtain high yield. CO_2 conversion to carbon monoxide (CO) looks like the simplest route for CO_2 reduction [121]. CO is a feedstock or intermediate product for the production of methanol and hydrocarbon fuels via Fischer-Tropsch synthesis of CH_4/CO_2 reforming to form syngas (CO/H_2) [122]. CO_2 reforming with CH_4 is an example of CO_2 being used as a soft oxidant, where the dioxide is dissociated into CO and surface oxygen, and oxygen abstracts hydrogen from methane to form water via the water-gas shift reaction (WGS) (Eqs. 11 and 12) [100-103, 121]. The catalytic chemistry of the reverse water-gas shift reaction and the following transformation to methanol/DME (or hydrocarbons via Fischer-Tropsch synthesis), and the subsequent production of gasoline (methanol-to-gasoline or diesel via hydrocracking of the alkanes produced in the Fisher-Tropsch process) are well established [102, 117-122]. On the other hand, methanol can be produced directly from carbon dioxide sources by catalytic hydrogenation and photo-assisted electrochemical reduction. A wide variety of CO_2 photo-reduction methods have been performed to oxygenate products, including formic acid (HCOOH) and formaldehyde (HCHO). HCOOH and

HCHO are the simplest oxygenates produced from the reduction of CO_2 with H_2O (or proton solvents) [121]. Furthermore, CO_2 can be utilized as a monomeric building block to synthesize various value-added oxygen-rich compounds and polymers under mild conditions. As an example, chemical conversion of CO_2 through C–N bond formation can produce value-added chemicals such as oxazolidinones, quinazolines, carbamates, isocyanates and polyurethanes [105]. These commodity chemicals have been synthesized from green methods and have important applications in the pharmaceutical and plastic industries. The chemisorption of CO_2 based on C–N bond formation could be one of the most efficient strategies, utilizing liquid absorbents such as conventional aqueous amine solutions, chilled ammonia, amino-functionalized ionic liquids, and solid absorbents including amino-functionalized silica, carbon, polymers and resins. The processes by which chemicals for CO_2 capture are manufactured should also be considered in terms of their energy requirements, efficiencies, waste products, and CO_2 emissions [105, 123]. In that sense, dimethyl carbonate (DMC) is a promising target molecule derived from CO_2 catalyzed by inorganic dehydrating agents such as molecular sieves [107]. Dimethyl carbonate has received much attention as a safe, non-corrosive, and environmentally friendly building block for the production of polycarbonates and other chemicals, an additive to fuel oil owing to its high octane number and an electrolyte in lithium batteries due to its high dielectric constant. It can be synthesized through a two-step transesterification process utilizing CO_2 as raw material [105, 107].

As a complementary technology to carbon sequestration and storage (CSS), the chemical recycling of carbon dioxide to fuels is an interesting opportunity. Chemical compounds such as alkane products ($CnH(2n+2)$) are un-branched hydrocarbons suitable for diesel fuel and jet fuel [121]. In this regard, biofuels or biodiesel, catalyzed using ceramic materials, can provide a significant contribution in energy independence and mitigation of climate change [109-127]. Today the main renewable biofuels are bioethanol and biodiesel. Biodiesel is a liquid fuel consisting of mono alkyl esters (methyl or ethyl) of long-chain fatty acids derived from vegetable oils, animal fats or micro and macro algal oils [127]. Biodiesel is a sustainable, renewable, non-toxic, biodegradable diesel fuel substitute that can be employed in current diesel engines without major modification, offering an interesting alternative to petroleum-based diesel [106, 111-115, 124-128]. Besides this, it is free from sulfur and aromatic components, making it cleaner burning than petroleum diesel. Biodiesel has a high flash point, better viscosity and caloric power similar to fossil fuels. It can be mixed with petroleum fossil fuel at any weight ratio or percentage, and it can be used without blending with fossil fuel (B100) as a successful fuel [127, 128]. It has similar properties (physical and chemical) to petroleum diesel fuel. Recently, transesterification (also called alcoholysis) has been reported as the most common way to produce biodiesel with lipid feedstock (such as vegetable oil or algal oil) and alcohol (usually methanol or ethanol), in presence of an acid or base catalyst. Transesterification is the best method for producing higher-quality biodiesel and glycerol [108, 110-115, 124-132]. The reaction is facilitated with a suitable catalyst [129-131]. The catalyst presence is necessary to increase both, the reaction rate and the transesterification reaction conversion yield. The catalysts are classified as homogeneous or heterogeneous. Homogeneous catalysts act in the same liquid phase as the reaction mixture. Conversely, if the catalyst remains in a different phase, the process is called heterogeneous catalytic transesterification [113, 127-131]. Heterogeneous catalysts are mostly applied in transesterification reaction due to many advantages such as easy catalyst separation and reusability, improved selectivity, fewer process stages, no water formation or saponification reaction, including in green technology, and cost effectiveness [127, 132]. The heterogeneous catalysts increase the mass transfer rate during the transesterification reaction [127, 131]. Various ceramic materials have been investigated for the production of biodiesel [106, 109-115, 124-179]. Some of these solid catalysts include alkali and alkaline-earth metal carbonates and oxides such as magnesium oxide (MgO), calcium oxide (CaO), barium oxide (BaO), strontium oxide (SrO) [124-131, 133-143]; lithium base ceramics (Li_4SiO_4 and Li_2SiO_3 [144-146]); sodium silicate (Na_2SiO_3 [147]); transition metal oxides and derivatives (titanium oxide, zinc oxide, mixed oxides catalysts [148-149]); ion exchange resin type acid heterogeneous catalysts [150]; MCM-metal impregnated materials [114]; layered double hydroxides (hydrotalcite-like hydroxides) [151-154]; hydrocalumite-like compounds [110,155]; supported bases [156-163]; and zeolites [164-165].

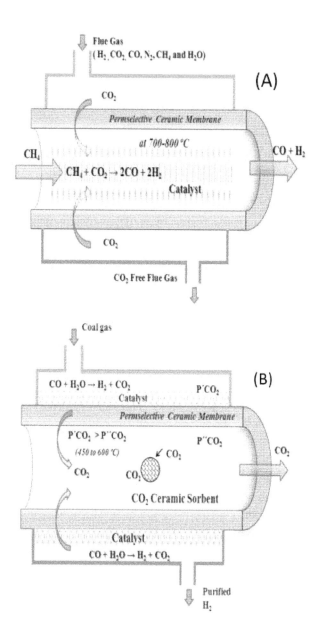

FIGURE 6: Schematic representation of the membrane reactor concept using a CO_2 permselective ceramic membrane: (a) CO_2 dry methane reforming and (b) water-gas shift reaction with hydrogen purification wherein CO_2 capture promotes the separation process.

Among the alkaline earth metal oxides, CaO is a promising basic heterogeneous catalyst for synthesizing biodiesel at mild temperatures (below the boiling point of methanol, MeOH) and at atmospheric pressure due to its plentiful availability and low cost, but it is rapidly hydrated and carbonated upon contact with room temperature air. CaO is the most widely used catalyst for transesterification and produces a high yield of 98% of fatty acid methyl esters (FAME) during the first cycle of reaction [130]. Granados et al. [142] used activated CaO as a solid base catalyst in the transesterification of sunflower oil to investigate the role of water and carbon dioxide on the deterioration of the catalytic performance upon contact with air for different periods. The study showed that CaO was rapidly hydrated and carbonated in air. Consequently, the reusability of the catalyst for subsequent steps is a big question mark. Di Serio et al. [170] reported a 92% biodiesel yield with MgO catalyst, using 12:1 methanol to oil molar ratio with 5.0wt% of the catalyst at methanol supercritical condition for 1 h. Wen et al. [171] carried out transesterification from waste cooking oil with methanol at 170 °C for 6 h with 10wt% of MgO/TiO2 and 50:1 M ratio of MeOH and oil. Guo et al. [172] studied the methyl ester yield produced via transesterification of soybean oil using sodium silicate as a catalyst. Sodium silicate was an effective catalyst for the microwave-irradiated production of biodiesel and hydrothermal production of hydrogen from by-product glycerol combined with Ni catalyst. The optimum reaction conditions obtained were 7.5:1 M ratio of alcohol/oil, 3wt% catalyst amount, 1 h reaction time and 60 °C reaction temperature. The FAME yield was ~100%. On the other hand, microwave-assisted transesterification of vegetable oil with sodium silicate is an effective and economical method for the rapid production of biodiesel. The reused catalyst after transesterification process for four cycles was recovered. Overall, sodium silicate was fully used in biodiesel production and glycerol gasification, and this co-production process provided a novel green method for biodiesel production and glycerol utilization [172].

Several techniques have been investigated for the transesterification reaction using heterogeneous catalysts for biodiesel production, as follows: transesterification via radio frequency microwaves, alcohol reflex temperature, alcohol supercritical temperature and ultrasonication [127,

173-177]. Recently, the use of ultrasonic irradiation has gained interest in biodiesel production [173-177]. Ultrasonic energy can emulsify the reactants to reduce the catalyst requirement, methanol-oil ratio, reaction time and reaction temperature and also provides the mechanical energy for mixing and the required activation energy for initiating the transesterification reaction [173-176]. The ultrasound phenomenon has its own physical and chemical effects on the liquid-liquid heterogeneous reaction system through cavitation bubbles, according to the following principles [175]: (1) the chemical effect, in which radicals such as H^+ and OH^- are produced during a transient implosive collapse of bubbles (in a liquid irradiated with ultrasound), which accelerates chemical reaction in the bulk medium; and (2) the physical effect of emulsification, in which the microturbulence generated due to radial motion of bubbles leads to intimate mixing (homogenizing the mixture) of the immiscible reactants. Accordingly, the interfacial region between the oil and alcohol increases sharply, resulting in faster reaction kinetics and higher conversion of oil and biodiesel yield [127]. In 2000, the ultrasonication reactor was first introduced by Hielscher Ultrasonic GmbH for biodiesel production. Nishimura et al. [175] studied the transesterification of vegetable oil using low-frequency ultrasound (28-40kHz). An excellent yield (~98%) was obtained at a 28 kHz ultrasound while a significant reduction of reaction time was obtained by using 40 kHz ultrasound. Salamatinia et al. [176] used ultrasonic assisted transesterification to improve the reaction rate. In this study, they used SrO and BaO as heterogeneous catalysts in the production of biodiesel from palm oil. The results showed that the basic properties of the catalyst were the main cause for their high activity. The low-frequency ultrasonic assisted transesterification process had no significant mechanical effects on SrO, but BaO catalyst study confirmed that the ultrasound treatment significantly improved the process by reducing the reaction time to less than 50 min at a catalyst loading of 2.8wt% to achieve biodiesel yield higher than 95%. Another study of alkali earth metals was carried out by Mootabadi et al. [177]. They reported the effect of ultrasonic waves at 20 kHz and 200W on the regenerated catalyst and compared mechanical stirring and ultrasonic irradiation. They investigated the optimum conditions, using palm oil for biodiesel production with catalysts such as CaO, SrO and BaO. They concluded that catalyst leaching was the main cause for the catalyst inactivity in the case of the re-used catalyst. BaO catalyst was found to be stable during the leaching. At the optimized condition, 95.2% yield was achieved with 60 min of reaction time for both BaO and SrO catalysts. For CaO catalyst, 77.3% yield was achieved with the same conditions. The use of ultrasound showed great enhancement of the reaction parameters in terms of the obtained yield and reaction time. The obtained yields were 30 to 40% higher in comparison to the corresponding results obtained using a conventional stirring reactor system without ultrasonication. Deng et al. [178] prepared nano-sized mixed Mg/Al oxides. Due to their strong basicity, the nanoparticles were further used as catalyst for biodiesel production from jatropha oil. Experiments were conducted with the solid basic catalyst in an ultrasonic transesterification reaction. Under the optimum conditions, biodiesel yield was 95.2%. After removing the glycerol on the catalyst surface, the nano-sized mixed Mg/Al oxides were reused eight times. The authors concluded that calcination of hydrotalcite nanocatalyst under ultrasonic radiation is an effective method for the production of biodiesel from jatropha oil. The activity of base solid catalysts is associated to their basic strength, such that the most basic catalyst showed the highest conversion. In another work, Deng et al. [179] reported optimum conditions for biodiesel production in the presence of base solid catalysts. They studied BaO and Ca–Mg–Al hydrotalcite (the most effective). The 95% biodiesel yield from jatropha oils and Ca–Mg–Al hydrotalcite was established with 30 min of reaction time. Ca–Mg–Al hydrotalcite could be reused twelve times after washing of the adsorbed glycerol from the catalyst surface with ethanol. Other types of heterogeneous catalysts under ultrasonic irradiation were used for transesterification by Georgogianni et al. [114]. They studied a wide range of catalysts including Mg-MCM-41, Mg–Al hydrotalcite and K^+-impregnated zirconium oxide. They mixed frying oils, methanol and the catalyst in a batch reactor with mechanical stirring for 24 h and with ultrasonication for 5 h. The results suggested that the basic strength was the cause of the good activity of the catalysts. Mg–Al hydrotalcite achieved the highest reaction conversion of 87% at a reaction temperature of 60 °C. Overall, ultrasonic irradiation significantly enhanced the reaction rate, causing a reduction in reaction time, and the biodiesel yield increased [114]. Consequently, a better understanding of the use of ultrasonic sound waves to accelerate the transesterification process could lead to substantial future improvement of both batch and continuous production systems, to obtain a more sustainable biodiesel production process [127].

REFERENCES

1. Ortiz-Landeros J.; Ávalos-Rendón T. L.; Gómez-Yáñez C.; Pfeiffer H. Analysis and Perspectives Concerning CO2 Chemisorption on Lithium Ceramics Using Thermal Analysis. J. Therm. Anal. Calorim. 2012, 108, 647–655.
2. Ávalos-Rendón T.; Casa-Madrid J.; Pfeiffer H. Thermochemical Capture of Carbon Dioxide on Lithium Aluminates (LiAlO2 and Li5AlO4): A New Option for the CO2 Absorption. J. Phys. Chem. A 2009, 113, 6919–6923.
3. Mejia-Trejo V. L.; Fregoso-Israel E.; Pfeiffer H. Textural, Structural, and CO2 Chemisorption Effects Produced on the Lithium Orthosilicate by Its Doping with Sodium (Li4−xNaxSiO4). Chem. Mater. 2008, 20, 7171–7176.
4. Mosqueda H. A.; Vazquez C.; Bosch P.; Pfeiffer H. Chemical Sorption of Carbon Dioxide (CO2) on Lithium Oxide (Li2O). Chem. Mater. 2006, 18, 2307–2310.
5. Shan S. Y.; Jia Q. M.; Jiang L. H.; Li Q. C.; Wang Y. M.; Peng J. H. Novel Li4SiO4-Based Sorbents from Diatomite for High Temperature CO2 Capture. Ceram. Int. 2013, 39, 5437–5441.
6. Olivares-Marín M.; Castro-Díaz M.; Drage T. C.; Maroto-Valer M. M. Use of Small-Amplitude Oscillatory Shear Rheometry to Study the Flow Properties of Pure and Potassium-Doped Li2ZrO3 Sorbents During the Sorption of CO2 at High Temperatures. Sep. Purif. Technol. 2010, 73, 415–420.
7. Pacciani R.; Torres J.; Solsona P.; Coe C.; Quinn R.; Hufton J.; Golden T.; Vega L. F. Influence of the Concentration of CO2 and SO2 on the Absorption of CO2 by a Lithium Orthosilicate-Based Absorbent. Environ. Sci. Technol. 2011, 45, 7083–7088.
8. Xiao Q.; Tang X.; Liu Y.; Zhong Y.; Zhu W. Citrate Route to Prepare K-Doped Li2ZrO3 Sorbents with Excellent CO2 Capture Properties. Chem. Eng. J. 2011, 174, 231–235.
9. Xiao Q.; Liu Y.; Zhong Y.; Zhu W. A Citrate Sol-Gel Method to Synthesize Li2ZrO3 Nanocrystals with Improved CO2 Capture Properties. J. Mater. Chem. 2011, 21, 3838–3842.
10. Rodríguez-Mosqueda R.; Pfeiffer H. Thermokinetic Analysis of the CO2 Chemisorption on Li4SiO4 by Using Different Gas Flow Rates and Particle Sizes. J. Phys. Chem. A 2010, 114, 4535–4541.
11. Ortiz-Landeros J.; Gomez-Yañez C.; Palacios-Romero L. M.; Lima E.; Pfeiffer H. Structural and Thermochemical Chemisorption of CO2 on Li4+x(Si1−xAlx)O4 and Li4−x(Si1−xVx)O4 Solid Solutions. J. Phys. Chem. A 2012, 116, 3163–3171.
12. Alcerreca-Corte I.; Fregoso-Israel E.; Pfeiffer H. CO2 Absorption on Na2ZrO3: A Kinetic Analysis of the Chemisorption and Diffusion Processes. J. Phys. Chem. C 2008, 112, 6520–6525.

13. Pfeiffer H.; Vazquez C.; Lara V. H.; Bosch P. Thermal Behavior and CO2 Absorption of Li2-xNaxZrO3 Solid Solutions. Chem. Mater. 2007, 19, 922−926.

14. Zhao T.; Ochoa-Fernández E.; Rønning M.; Chen D. Preparation and High-Temperature CO2 Capture Properties of Nanocrystalline Na2ZrO3. Chem. Mater. 2007, 19, 3294−3301.

15. Seggiani M.; Puccini M.; Vitolo S. Alkali Promoted Lithium Orthosilicate for CO2 Capture at High Temperature and Low Concentration. Int. J. Greenhouse Gas Control 2013, 17, 25−31.

16. Khokhani M.; Khomane R. B.; Kulkarni B. D. Sodium-Doped Lithium Zirconate Nano-Squares: Synthesis, Characterization and Applications for CO2 Sequestration. J. Sol-Gel Sci. Technol. 2012, 61, 316−320.

17. Veliz-Enriquez M. Y.; Gonzalez G.; Pfeiffer H. Synthesis and CO2 Capture Evaluation of Li2-xKxZrO3 Solid Solutions and Crystal Structure of a New Lithium-Potassium Zirconate Phase. J. Solid State Chem. 2007, 180, 2485−2492.

18. Martínez-dlCruz L.; Pfeiffer H. Microstructural Thermal Evolution of the Na2CO3 Phase Produced During a Na2ZrO3−CO2 Chemisorption Process. J. Phys. Chem. 2012, 116, 9675−9680.

19. Santillan-Reyes G. G.; Pfeiffer H. Analysis of the CO2 Capture in Sodium Zirconate (Na2ZrO3). Effect of the Water Vapor Addition. Int. J. Greenhouse Gas Control 2011, 5, 1624−1629.

20. Iwana A.; Stephenson H.; Ketchie C.; Lapkin A. High Temperature Sequestration of CO2 Using Lithium Zirconates. Chem. Eng. J. 2009, 146, 249−258.

21. Tabarés F. L. Editor, Lithium: Technology, Performance and Safety, Nova Publishers, (2013). Chapter 6, Lithium Ceramics as an Alternative for the CO2 Capture. Analysis of Different Physicochemical Factors Controlling this Process, pp 171-192.

22. Bish D.L., Anion-Exchange in Takovite: Applications to Other Hydroxide Minerals, Bone Miner., 1980, 103, 170-175.

23. Duan X.; Evans D. G., Layered Double Hydroxides. Structure and Bonding, Eds. Springer-Verlag: Berlin Heidelberg. Germany, 2006; vol. 119.

24. Wang M. Z.; Hu Q. D. L.; Li Y.; Li S.; Zhang X.; Xi M.; Yang X., Intercalation of Ga3+-Salicylidene-Amino Acid Schiff Base Complexes into Layered Double Hydroxides: Synthesis, Characterization, Acid Resistant Property, in Vitro Release Kinetics and Antimicrobial Activity, Appl. Clay Sci. 2013, 83&84, 182-190.

25. Catti M.; Ferraris G.; Hull S.; Pavese A. Static Compression and H Disorder in Mg(OH)2 (Brucite) to 11 GPa: a Powder Neutron Diffraction Study. Phys. Chem. Miner. 1995, 22, 200-206.

26. He J.; Wei M.; Li B.; Kang Y.; Evans D. G.; Duan X., Preparation of Layered Double Hydroxides, Structure and Bonding, 2005, 119, 89-119.

27. Gutmann N.; Müller B. Insertion of the Dinuclear Dihydroxo-Bridged Cr(III) Aquo Complex into the Layered Double Hydroxides of Hydrotalcite-Type. J. Solid State Chem. 1996, 122, 214-220.

28. Fogg A.M.; Williams G.R.; Chester R.; O'Hare D. A Novel Family of Layered Double Hydroxides—[MgAl4(OH)12] (NO3)2 xH2O (M = Co, Ni, Cu, Zn). J. Mater. Chem., 2004, 14, 2369-2371.

29. de Roy A.; Forano C.; Besse J.P. Layered Double Hydroxides: Synthesis and Post-Synthesis Modification. Review, Chapter 1, 2002.

30. Choi S.; Drese J.H.; Jones C.W. Adsorbent Materials for Carbon Dioxide Capture from Large Anthropogenic Point Sources. ChemSusChem. 2009, 2, 796-854.

31. Ding Y.; Alpay, E. Equilibria and Kinetics of CO2 Adsorption on Hydrotalcite Adsorbent. Chem. Eng. Sci. 2000, 55, 3461-3474.

32. Reijers H.T.J.; Valster-Schiermeier S.E.A.; Cobden P.D.; van der Brink R.W. Hydrotalcite as CO2 Sorbent for Sorption-Enhanced Steam Reforming of Methane. Ind. Eng. Chem. Res. 2006, 45, 2522-2530.

33. Wang X.P.; Yu J.J.; Cheng J.; Hao Z.P.; Xu Z.P. High Temperature Adsorption of Carbon Dioxide on Mixed Oxides Derived from Hydrotalcite-Like Compounds. Environ. Sci. Technol. 2008, 42, 614-618.

34. Reynolds S.P.; Ebner A.D.; Ritter J.A. Carbon Dioxide Capture from Flue Gas by Pressure Swing Adsorption at High Temperature Using a K-Promoted HTLC: Effects of Mass Transfer on the Process Performance. Environ. Prog. 2006, 25, 334-342.

35. Reynolds, S.P.; Ebner, A.D.; Ritter, J.A. Stripping PSA Cycles for CO2 Recovery from Flue Gas at High Temperature Using a Hydrotalcite Like Adsorbent. Ind. Eng. Chem. Res. 2006, 45, 4278-4294.

36. Ebner, A.D.; Reynolds, S.P.; Ritter, J.A. Nonequilibrium Kinetic Model that Describes the Reversible Adsorption and Desorption Behavior of CO2 in a K-Promoted Hydrotalcite-Like Compound. Ind. Eng. Chem. Res. 2007, 46, 1737-1744.

37. Cavani F.; Trifirb F.; Vaccari A. Hydrotalcite-Type Anionic Clays: Preparation, Properties and Applications, Catal. Today, 1991, 11, 173–301.

38. Vaccari A. Preparation and Catalytic Properties of Cationic and Anionic Clays, Catal. Today, 1998, 41, 53–71.

39. Das N.N.; Konar J.; Mohanta M.K.; Srivastava S.C. Adsorption of Cr(VI) and Se(IV) from their Aqueous Solutions onto Zr4+-Substituted ZnAl/MgAl-Layered Double Hydroxides: Effect of Zr4+ Substitution in the Layer, J. Colloid Interf. Sci., 2004, 270, 1–8.

40. Goh K.H.; Lim T.T; Dong Z. Application of Layered Double Hydroxides for Removal of Oxyanions: A Review, Water Research, 2008, 42, 1343–1368.

41. Yong Z.; Mata, V.; Rodriguez, A.E. Adsorption of Carbon Dioxide onto Hydrotalcite-Like Compounds (HTLCs) at High Temperatures. Ind. Eng. Chem. Res. 2001, 40, 204-209.

42. Bellotto M.; Rebours B.; Clause O.; Lynch J.; Bazin D.; Elkaîm E. Hydrotalcite Decomposition Mechanism: A Clue to the Structure and Reactivity of Spinel-Like Mixed Oxides, J. Phys. Chem. 1996, 100, 8535-8542.

43. Ram Reddy M. K., Xu Z. P., Lu G. Q. (Max); Diniz da Costa J. C. Layered Double Hydroxides for CO2 Capture: Structure Evolution And Regeneration. Ind. Eng. Chem. Res. 2006, 45, 7504-7509.

44. Hufton J.; Mayorga S.; Gaffeney T.; Nataraj S.; Sircar S. Sorption Enhanced Reaction Process (SERP), Proceedings of the 1997 U.S., DOE Hydrogen Program Review, 1997, 1, 179-194.

45. Ram Reddy M.K.; Xu Z.P.; Lu G.Q. (Max); Diniz da Costa J.C. Influence of Water on High-Temperature CO2 Capture Using Layered Double Hydroxide Derivatives. Ind. Eng. Chem. Res. 2008, 47, 2630-2635.

46. Martunus; Othman M.R; Fernando W.J.N. Elevated Temperature Carbon Dioxide Capture Via Reinforced Metal Hydrotalcite. Micropor. Mesopor. Mater. 2011, 138, 110–117.

47. Yong Z.; Rodrigues A.E. Hydrotalcite-Like Compounds as Adsorbents for Carbon Dioxide. Energy Convers. & Manage. 2002, 43, 1865-1876.

48. Newman S. P.; Jones W. Supramolecular Organization and Materials Design, Cambridge University Press, England, 2001.

49. Yamamoto T.; Kodama T.; Hasegawa N.; Tsuji M.; Tamura Y. Synthesis of Hydrotalcite with High Layer Charge for CO2 Adsorbent. Energy Convers Mgmt, 1995, 36, 637-640.

50. Wang Q.; Wu Z.; Tay H. H.; Chen L.; Liu Y.; Chang J.; Zhong Z.; Luo J.; Borgna A. High Temperature Adsorption of CO2 on Mg-Al Hydrotalcite: Effect of the Charge Compensating Anions and the Synthesis pH. Catal. Today, 2011, 164, 198-203

51. Wang Q.; Tay H.H.; Ng D.J.W.; Chen L.; Liu Y.; Chang J.; Zhong Z.; Luo J.; Borgna A. The Effect of Trivalent Cations on the Performance of Mg-M-CO3 Layered Double Hydroxides for High-Temperature CO2 Capture. ChemSusChem. 2010, 3, 965-973.

52. Hibino T.; Yamashita Y.; Kosuge K.; Tsunashima A. Decarbonation Behavior of Mg-Al-CO3 Hydrotalcite-Like Compounds During Heat Treatment,. Clays Clay Minerals. 1995, 43, 427 – 432.

53. Qian W.; Luo J.; Zhong Z.; Borgna A. CO2 Capture by Solid Adsorbents and their Applications: Current Status and New Trends. Energy Environ. Sci., 2011, 4, 42-55.

54. Hufton J. R.; Mayorga S.; Sircar S. Sorption-Enhanced Reaction Process for Hydrogen Production. AIChE J. 1999, 45, 248.

55. Oliveira E.L.G.; Grande C.A.; Rodrigues A.E.; CO2 Sorption on Hydrotalcite and Alkali-Modified (K and Cs) Hydrotalcites at High Temperatures. Sep. Purif. Technol. 2008, 62, 137-147.

56. Yang J. I.; Kim J. N. Hydrotalcites for Adsorption of CO2 at High Temperature. Korean J. Chem. Eng., 2006, 23, 77-80.

57. Ida J.I.; Lin S. Mechanism of High-Temperature CO2 Sorption on Lithium Zirconate. Environ. Sci. Technol., 2003, 37, 1999-2004.

58. Hufton J.; Mayorga S.; Nataraj S.; Sircar S.; Rao M. Sorption-Enhanced Reaction Process (SERP), Proceedings of the 1998, USDOE Hydrogen Program Review, 1998, 2, 693-705.

59. Lee Jung M.; Min Yoon J.; Lee Ki B.; Jeon Sang G.; Na Jeong G.; Ryu Ho J. Enhancement of CO2 Sorption Uptake on Hydrotalcite by Impregnation with K2CO3. Langmuir, 2010, 26, 18788–18797.

60. Baker R. W. Membrane Technology and Applications, 2nd Editions, John Wiley and Sons (2004).

61. Murder M., Basic Principles of Membrane Technology, Kluwer Academic Publishers (1991).

62. Aresta M.; Dibenedetto A. The Contribution of the Utilization Option to Reducing the CO2 Atmospheric Loading: Research Needed to Overcome Existing Barriers for a Full Exploitation of the Potential of The CO2 Use, Cat. Today. 2004, 98, 455–462.

63. Anderson M.; Wang H.; Lin Y. S. Inorganic Membranes for Carbon Dioxide and Nitrogen Separation, Rev. Chem. Eng., 2012, 28, 101-121.

64. Yang H.; Xu Z.; Fan M.; Gupta R.; Slimane R. B.; Bland A. E.; Wright I. Progress in Carbon Dioxide Separation and Capture: A Review, J. Environ. Sci., 2008, 20, 14-27.
65. Niwa M.; Katada N.; Okumura K. Introduction to Zeolite Science and Catalysis, Characterization and Design of Zeolite Catalysts, Springer Series in Materials Science, Vol. 141 (2010).
66. Iwamoto Y.; Kawamoto H. Science and Technology Trends: Quarterly Report, 2009, 32, 42-59.
67. Algieri C.; Barbieri G.; Drioli E.; Zeolite Membranes for Gas Separations, in Membrane Engineering for the Treatment of Gases. Royal Society of Chemistry Vol. 2 (2011).
68. Fedosov D. A.; Smirnov A. V.; Knyazeva E. E.; Ivanova I. I. Zeolite membranes: Synthesis, properties, and application, Petroleum Chem., 2011, 51, 657-667.
69. Caro J.; Noack M. Zeolite membranes – Recent developments and progress. Micropor. Mesopor. Mater. 2008, 115, 215–233.
70. Jia M. D.; Peinemann K. V.; Behling R. D. Ceramic Zeolite Composite Membranes. Preparation, Characterization and Gas Permeation. J. Memb. Sci. 1993, 82, 15-26.
71. Cui Y.; Kita H.; Okamoto K. I. Preparation and Gas Separation Performance of Zeolite T Membrane. J. Mater. Chem. 2004, 14, 924.
72. Poshusta J.; Tuan V.; Pape E.; Noble R.; Falconer J. Separation of Light Gas Mixtures Using SAPO-34 Membranes. AIChE J. 2000, 46, 779-789.
73. Li S.; Falconer J.; Noble R. SAPO-34 Membranes for CO2/CH4 Separation. J. Memb. Sci., 2004, 241, 121–135.
74. Tomita T.; Nakayama K.; Sakai H. Gas Separation Characteristics of DDR Type Zeolite Membrane. Micropor. Mesopor. Mater. 2004, 68, 71.
75. Caro J.; Noack M. Zeolite Membranes: Recent Developments and Progress. Micropor. Mesopor. Mater. 2008, 115, 215-233.
76. Burggraaf A. J. Fundamentals of Inorganic Membrane Science and Technology, Membrane Science and Technology, Series, 4, Elsevier Science. Netherlands (1996).
77. Lara-Medina J. J.; Torres-Rodriguez M.; Gutierrez-Arzaluz M.; Mugica-Alvarez V. Separation of CO2 and N2 with a Lithium-Modified Silicalite-1 Zeolite Membrane. Inter. J. Greenhouse Gas Control, 2012, 10, 494-500.
78. Dyer A. Introduction to Zeolite Science and Practice. 3rd Revised Edition. J. Cejka, H. van Bekkum, A. Corma and F. Schüth (Editors) Elsevier B.V. (2007).
79. An W.; Swenson P.; Gupta A.; Wu L.; Kuznicki T. M.; Kuznicki S. M. Improvement of H2/CO2 Selectivity of the Natural Clinoptilolite Membranes by Cation Exchange Modification. J. Memb. Sci., 2013, 433, 25–31.
80. White J. C.; Dutta P. K.; Shqau K.; Verweij H. Synthesis of Ultrathin Zeolite Y Membranes and their Application for Separation of Carbon Dioxide and Nitrogen Gases. Langmuir 2010, 26, 12, 10287–10293.
81. Cho Y.K., Han K., Lee K.H. Separation of CO2 by Modified γ-Al2O3 Membranes at High Temperature. J. Membrane Sci. 1995, 104, 219-230.
82. Keizer K.; Uhlhorn R.J.R.; van Vuren R.J.; Burggraaf A.J. Gas Separation Mechanisms in Microporous Modified γ-A12O3 Membranes. J. Membrane Sci., 1988, 39, 285-300.

83. Uhlhorn R.J.R.; Keizer K., Burggraaf A.J. Gas and Surface Diffusion in Modified 33-Alumina Systems. J. Membrane Sci., 1989, 46, 225-241.
84. Kusakabe K., Ichiki K., Morooka S. Separation of CO_2 with $BaTiO_3$ Membrane Prepared by the Sol—Gel Method. J. Membrane Sci. 1944, 95, 171-177.
85. Nomura M., Sakanishi T., Nishi Y., Utsumi K., Nakamura R.. Preparation of CO_2 Permselective Li_4SiO_4 Membranes by Using Mesoporous Silica as a Silica Source. Energy Procedia 2013, 37, 1004-1011.
86. Nomura M.; Nishi Y.; Sakanishi T.; Utsumi K.; Nakamura R. Preparation of Thin Li-4SiO_4 Membranes by Using a CVD Method, Energy Procedia 2013, 37, 1012-1019.
87. Kawamura H.; Yamaguchi T.; Nair B. N.; Nakagawa K.; Nakao S.I. Dual-Ion Conducting Lithium Zirconate-Based Membranes for High Temperature CO_2 Separation. J. Chem. Eng. Jpn. 2005, 38, (5) 322-328.
88. Yamaguchi T.; Niitsume T.; Nair B. N.; Nakagawa K. Lithium Silicate Based Membranes for High Temperature CO_2 Separation. J. of Membrane Sci., 2007, 294, 16-21.
89. Nair B.N.; Burwood R.P.; Goh V.J.; Nakagawa K.; Yamaguchi T. Lithium Based Ceramic Materials and Membranes for High Temperature CO_2 Separation. Progress in Materials Sci. 2009, 54, 511–541.
90. Skinner S.J.; Kilner J.A. Oxygen Ion Conductors. Materials Today, 2003, 6, 30-37.
91. Anderson M.; Lin Y.S. Carbonate–Ceramic Dual-Phase Membrane for Carbon Dioxide Separation. J. Membrane Sci. 2010, 357, 22.
92. Wade J. L.; Lee C.; West A. C.; Lackner K. S. Composite Electrolyte Membranes for High Temperature CO_2 Separation. J. Membrane Sci. 2011, 369, 20.
93. Zhang L.; Xu N.; Li X.; Wang S.; Huang K.; Harris W. H.; Wilson K.; Chiu S. High CO_2 Permeation Flux Enabled by Highly Interconnected Three-Dimensional Ionic Channels in Selective CO_2 Separation Membranes. Energy Environ. Sci. 2012, 5, 8310.
94. Rui Z.; Anderson M.; Li Y.; Lin Y.S. Ionic Conducting Ceramic and Carbonate Dual Phase Membranes for Carbon Dioxide Separation, J. Membrane Sci. 2012, 417-418, 174.
95. Lu B.; Lin Y.S. Synthesis and Characterization of Thin Ceramic-Carbonate Dual-Phase Membranes for Carbon Dioxide Separation. J. Membrane Sci. 2013, 444, 402–411.
96. Dong X.; Ortiz-Landeros J.; Lin Y. S. An Assymetric Thin Tubular Dual Phase Membrane. Chem. Commun. 2013, 49, 9654.
97. Ortiz-Landeros J.; Norton T.T.; Lin Y.S. Effects of Support Pore Structure on Carbon Dioxide Permeation of Ceramic-Carbonate Dual-Phase Membranes. Chem. Eng. Sci. 2013, 104, 891-898.
98. Rui Z.; Anderson M.; Lin Y.S.; Li Y.; Modeling and Analysis of Carbon Dioxide Permeation through Ceramic-Carbonate Dual-Phase Membranes. J. Membrane Sci. 2009, 345, 110.
99. Yu K.M.K.; Curcic I.; Gabriel J.; Tsang S.C.E. Recent Advances in CO_2 Capture and Utilization, ChemSusChem. 2008, 1, 893–899.
100. Wender I. Reactions of Synthesis Gas, Fuel Process. Technol. 1996, 48, 3, 189-297.
101. Rostrup-Nielsen J. R. Syngas in Perspective. Catal. Today 2002, 71, 3-4, 243-247.

102. Dry M.E. The Fischer–Tropsch Process: 1950–2000. Catal. Today 2002, 71, 227-241.

103. Yu K.M.K.; Curcic I.; Gabriel J.; Tsang S.C.E. Recent Advances in CO2 Capture and Utilization. ChemSusChem. 2008, 1, 893 – 899.

104. Gnanapragasam N.; Reddy B.; Rosen M. Reducing CO2 Emissions for an IGCC Power Generation System: Effect of Variations in Gasifier and System Operating Conditions. Energ. Convers. Manage. 2009, 50, 1915-1923.

105. Yang Z.Z.; He L.N.; Gao J.; Liu A.H.; Yu B. Carbon Dioxide Utilization with C–N Bond Formation: Carbon Dioxide Capture and Subsequent Conversion. Energy Environ. Sci. 2012, 5, 6602-6639.

106. Chattopadhyay S.; Sen R. Fuel Properties, Engine Performance and Environmental Benefits of Biodiesel Produced by a Green Process. Appl. Energ. 2013, 105, 319–326.

107. Choi J.C.; He L.N.; Yasuda H.; Sakakura T. Selective and High Yield Synthesis of Dimethyl Carbonate Directly from Carbon Dioxide and Methanol. Green Chem. 2002, 4, 230–234.

108. Quispe C.A.; Coronado J.R.C; CarvalhoJr J.A. Glycerol: Production, Consumption, Prices, Characterization and New Trends in Combustion. Renew. Sust. Energ. Rev. 2013, 27, 475–493.

109. Talebian-Kiakalaieh A.; Saidina Amin N. A; Mazaheri H. A Review on Novel Processes of Biodiesel Production from Waste Cooking Oil. Appl. Energ. 2013, 104, 683–710.

110. Kuwahara Y.; Tsuji K.; Ohmichi T.; Kamegawa T.; Moria K.; Yamashita H. Transesterifications Using a Hydrocalumite Synthesized from Waste Slag: An Economical and Ecological Route for Biofuel Production. Catal. Sci. Tech. 2012, 2, 1842–1851.

111. Atadashi I.M.; Aroua M.K.; Abdul Aziz A. Biodiesel Separation and Purification: A Review. Renew. Energ. 2011, 36, 437-443.

112. Lam M.K.; Lee K.T. Mixed Methanol–Ethanol Technology to Produce Greener Biodiesel from Waste Cooking Oil: A Breakthrough for $SO_4^{2-}/SnO_2–SiO_2$ catalyst. Fuel Process. Technol. 2011, 92, 1639–1645.

113. Lam M.; Lee K.T.; Rahman Mohamed A. Homogeneous, Heterogeneous and Enzymatic Catalysis for Transesterification of High Free Fatty Acid Oil (Waste Cooking Oil) to Biodiesel: A Review. Biotechnol. Adv. 2010, 28, 500–518.

114. Georgogianni K.G.; Katsoulidis A.K.; Pomonis P.J.; Manos G.; Kontominas M.G. Transesterification of Rapeseed Oil for the Production of Biodiesel Using Homogeneous and Heterogeneous Catalysis. Fuel Process. Technol. 2009, 90, 1016–1022.

115. Zhang Y.; Dube M.A.; McLean D.D.; Kates M. Biodiesel Production from Waste Cooking Oil: 1. Process Design and Technological Assessment. Bioresource Technol. 2003, 89, 1-16.

116. Long Y.D.; Fang Z.; Su T.C.; Yang Q. Co-production of Biodiesel and Hydrogen from Rapeseed and Jatropha Oils with Sodium Silicate and Ni Catalysts. Appl. Energ. 2013, http://dx.doi.org/10.1016/j.apenergy.2012.12.076.

117. Centi G.; Quadrelli E.A.; Perathoner S. Catalysis for CO2 Conversion: A Key Technology for Rapid Introduction of Renewable Energy in the Value Chain of Chemical Industries. Energy Environ. Sci. 2013, 6, 1711-1731.

118. Centi G.; Perathoner S. Opportunities and Prospects in the Chemical Recycling of Carbon Dioxide to Fuels. Catal. Today 2009, 148, 191-205.
119. Hoffmann F.M.; Yang Y.; Paul J.; White M.G; Hrbek J. Hydrogenation of Carbon Dioxide by Water: Alkali-Promoted Synthesis of Formate. J. Phys. Chem. Lett. 2010, 1, 2130–2134.
120. H. Yin, X. Mao, D. Tang, W. Xiao, L. Xing, H. Zhu, D. Wang, D.R. Sadoway. Capture and Electrochemical Conversion of CO2 to Value-Added Carbon and Oxygen by Molten Salt Electrolysis. Energy Environ. Sci., 2013, 6, 1538-1545.
121. Hu B.; Guild C.; Suib S.L. Thermal, Electrochemical, and Photochemical Conversion of CO2 to Fuels and Value-Added Products. Journal of CO2 Utilization. 2013, 1, 18–27.
122. Kumar B.; Smieja J.M.; Kubiak C.P. Photo-reduction of CO2 on p-type Silicon Using Re(Bipy-But)(CO)3Cl: Photovoltages Exceeding 600 mv for the Selective Reduction of CO2 to CO. J. Phys. Chem. C 2010, 114, 14220–14223.
123. Bara J. Review: The Chemistry of Amine Manufacture. What Chemicals Will We Need to Capture CO2? Greenhouse Gas Sci. Technol. 2012, 2, 162–171.
124. Helwani Z.; Othman M.R.; Aziz N.; Fernando W.J.N.; Kim J. Technology for Production of Biodiesel Focusing on Green Catalytic Techniques: A Review. Fuel Process. Technol. 2009, 90, 1502–1515.
125. Endalew K.; Kiros Y., Zanzi R. Inorganic Heterogeneous Catalysts for Biodiesel Production from Vegetable Oils. Biomass Bioenerg. 2011, 35, 3787-3809.
126. Luque R.; Lovett J.C.; Datta B.; Clancy J.; Campeloa J.M.; Romero A.A. Biodiesel as Feasible Petrol Fuel Replacement: A Multidisciplinary Overview. Energy Environ. Sci. 2010, 3, 1706–1721.
127. Ramachandran K.; Suganya T.; Gandhi N.N.; Renganathan S. Recent Developments for Biodiesel Production by Ultrasonic Assist Transesterification Using Different Heterogeneous Catalyst: A Review. Renew. Sust. Energ. Rev. 2013, 22, 410–418.
128. Vyas A.P.; Verma J.L.; Subrahmanyam N. A Review on FAME Production Processes. Fuel 2010, 8, 1–9.
129. Di Serio M., Ledda M., Cozzolino M., Minutillo G., Tesser R., Santacesaria E. Transesterification of Soybean Oil to Biodiesel by Using Heterogeneous Basic Catalysts. Ind. Eng. Chem. Res. 2006, 45, 3009-3014.
130. Veljkovic V. B.; Stamenkovic O. S.; Todorovic Z. B.; Lazic M. L.; Skala D. U. Kinetics of Sunflower Oil Methanolysis Catalyzed by Calcium Oxide. Fuel 2009, 88, 554–1562.
131. Singh Chouhan A.P.; Sarma A.K. Modern Heterogeneous Catalysts for Biodiesel Production: A Comprehensive Review. Renew. Sust. Energ. Rev. 2011, 15, 4378–4399.
132. Lee J.S.; Saka S. Biodiesel Production by Heterogeneous Catalysts and Super-Critical Technologies: Review. Bioresource Technol. 2010, 101, 7191–7200.
133. Salamatinia B.; Abdullah A.Z.; Bhatia S. Quality Evaluation of Biodiesel Produced through Ultrasound-Assisted Heterogeneous Catalytic System. Fuel Process. Technol. 2012, 97, 1-8.
134. Berrios M.; Martín M.A.; Chica A.F.; Martín A. Purification of Biodiesel from Used Cooking Oils. Appl. Energ. 2011, 88, 3625–3631.

135. Zabeti M.; Daud W.M.A.W.; Aroua M.K. Activity of Solid Catalysts for Biodiesel Production: A Review. Fuel Process. Technol. 2009, 90, 770–777.

136. Sharma Y. C.; Singh B.; Korstad J. Latest Developments on Application of Heterogenous Basic Catalysts for an Efficient and Eco Friendly Synthesis of Biodiesel: A Review. Fuel 2011, 90, 1309–1324.

137. Wen Z.; Yu X.; Tu S.T.; Yan J.; Dahlquist E. Synthesis of Biodiesel from Vegetable Oil with Methanol Catalyzed by Li-Doped Magnesium Oxide Catalysts. Appl. Energ. 2010, 87, 743–748.

138. Dossin T.F.; Reyniers M.F.; Berger R.J.; Marin G.B. Simulation of Heterogeneously MgO-Catalyzed Transesterification for Fine-Chemical and Biodiesel Industrial Production. Appl. Catal. B-Environ. 2006, 67, 136–148.

139. Liu X.; He H.; Wang Y.; Zhu S.; Piao X. Transesterification of Soybean Oil to Biodiesel Using CaO as a Solid Base Catalyst. Fuel 2008, 87, 216–21.

140. Liu X.; He H.; Wang Y.; Zhu S. Transesterification of Soybean Oil to Biodiesel Using SrO as a Solid Base Catalyst. Catal. Commun. 2007, 8, 1107–1111.

141. Soares Días A.P.; Bernardo J.; Felizardo P.; Neiva Correia M.J. Biodiesel Production by Soybean Oil Methanolysis over SrO/MgO Catalysts. The Relevance of the Catalyst Granulometry. Fuel Process. Technol. 2012, 102, 146–155.

142. Granados M.L.; Poves M.D.; Alonso D.; Mariscal R.; Galisteo F.C.; Moreno-Tost R. Biodiesel from Sunflower Oil by Using Activated Calcium Oxide. Appl. Catal. B-Environ. 2007, 73, 317–26.

143. Watkins R.S.; Lee A.F.; Wilson K. Li–CaO Catalysed Tri-Glyceride Transesterification for Biodiesel Applications. Green Chem. 2004, 6, 335–340.

144. Wang J.X.; Chen K.T.; Huang J.S.; Chen C.C. Application of Li2SiO3 as a Heterogeneous Catalyst in the Production of Biodiesel from Soybean Oil. Chinese Chem. Lett. 2011, 22, 1363–1366.

145. Wang J.X.; Chen K.T.; Wu J.S.; Wang P.H.; Huang S.T.; Chen C.C. Production of Biodiesel through Transesterification of Soybean Oil Using Lithium Orthosilicate Solid Catalyst. Fuel Process. Technol. 2012, 104, 167–173.

146. Chen K.T.; Wang J.X.; Dai Y.M.; Wang P.H.; Liou C.Y.; Nien C.W.; Wu J.S.; Chen C.C. Rice Husk Ash as a Catalyst Precursor for Biodiesel Production. J. Taiwan Inst. Chem. Eng. 2013, 44, 622-629.

147. Guo P.; Zheng C.; Zheng M.; Huang F.; Li W.; Huang Q. Solid Base Catalysts for Production of Fatty Acid Methyl Esters. Renew. Energ. 2013, 53, 377-383.

148. Singh A.K.; Fernando S.D. Transesterification of Soyabean Oil Using Heterogenous Catalysts. Energ. Fuels 2008, 22, 2067-2069.

149. Omar W.N.N.W.; Amin N.A.S. Biodiesel Production from Waste Cooking Oil over Alkaline Modified Zirconia Catalyst. Fuel Process. Technol. 2011, 92, 2397-2405.

150. Molaei Dehkordi A.; Ghasemi M. Transesterification of Waste Cooking Oil to Biodiesel Using Ca and Zr Mixed Oxides as Heterogeneous Base Catalysts. Fuel Process. Technol. 2012, 97, 45–51.

151. Shibasaki-Kitakawa N., Honda H., Kuribayashi H., Toda T., Fukumura T., Yonemoto T. Biodiesel Production Using Anionic Ion-Exchange Resin as Heterogeneous Catalyst. Bioresource Technol. 2007, 98, 416–421.

152. Shumaker J. L.; Crofcheck C.; Tackett S.A.; Santillan-Jimenez E.; Morgan T.; Ji Y.; Mark Crocker; Toops T.J. Biodiesel Synthesis Using Calcined Layered Double Hydroxide Catalysts. Appl. Catal. B-Environ. 2008, 82, 120–130.

153. Corma A., Hamid S.B.A., Iborra S., Velty A. Lewis and Bronsted Basic Active Sites on Solid Catalysts and their Role in the Synthesis of Monoglycerides. J. Catal. 2005, 234, 340-347.

154. Sankaranarayanan S.; Churchil Antonyraj A.; Kannan S. Transesterification of Edible, Non-Edible and used Cooking Oils for Biodiesel Production Using Calcined Layered Double Hydroxides as Reusable Base Catalysts. Bioresource Technol. 2012, 109, 57–62.

155. Navajas A.; Campo I.; Arzamendi G.; Hernandez W.Y.; Bobadilla L.F.; Centeno M.A.; Odriozola J.A.; Gandia L.M. Synthesis of Biodiesel from the Methanolysis of Sunflower Oil Using PURAL® Mg–Al Hydrotalcites as Catalyst Precursors. Appl. Catal. B-Environ 2010, 100, 299–309.

156. Gomes J.F.P.; Puna J.F.B.; Gonçalves L.M.; Bordado J.C.M. Study on the use of MgAl Hydrotalcites as Solid Heterogeneous Catalysts for Biodiesel Production. Energy 2011, 36, 6770-6778.

157. Sánchez-Cantú M.; Pérez-Díaz L.M.; Tepale-Ochoa N.; González-Coronel V.J.; Ramos-Cassellis M.E.; Machorro-Aguirre D.; Valente J.S. Green Synthesis of Hydrocalumite-Type Compounds and their Evaluation in the Transesterification of Castor Bean Oil and Methanol. Fuel 2013, 111, 23-31.

158. Sun H.; Ding Y.; Duan J.; Zhang Q.; Wang Z.; Lou H.; Zheng X. Transesterification of Sunflower Oil to Biodiesel on ZrO2 Supported La2O3 Catalyst. Bioresource Technol. 2010, 101, 953–958.

159. Kim H.J.; Kang B.S.; Kim M.J.; Park Y.M.; Kim D.K.; Lee J.S.; Lee K.Y. Transesterification of Vegetable Oil to Biodiesel Using Heterogeneous Base Catalyst. Catal. Today 2004, 93–95, 315–320.

160. Ebiura T.; Echizen T.; Ishikawa A.; Murai K.; Baba T. Selective Transesterification of Triolein with Methanol to Methyl Oleate and Glycerol Using Alumina Loaded with Alkali Metal Salt as a Solid-Base Catalyst. Appl. Catal. A-Gen. 2005, 283, 111–116.

161. Xie W.; Li H. Alumina-Supported Potassium Iodide as a Heterogeneous Catalyst for Biodiesel Production from Soybean Oil. J. Mol. Catal. A-Chem. 2006, 255, 1–9.

162. Lukic I.; Krstic J.; Jovanovic D.; Skala D. Alumina/Silica Supported K2CO3 as a Catalyst for Biodiesel Synthesis from Sunflower Oil. Bioresource Technol. 2009, 100, 4690–4696.

163. Evangelista J.P.C.; Chellappa T.; Coriolano A.C.F.; Fernandes Jr. V.J.; Souza L.D.; Araujo A.S.; Synthesis of Alumina Impregnated with Potassium Iodide Catalyst for Biodiesel Production from Rice Bran Oil. Fuel Process. Technol. 2012, 104, 90–95.

164. Arzamendi G.; Campo I.; Arguiñarena E.; Sánchez M.; Montes M.; Gandía L.M. Synthesis of Biodiesel with Heterogeneous NaOH/Alumina Catalysts: Comparison with Homogeneous NaOH. Chem. Eng. J. 2007, 134, 123–130.

165. Baroutian S.; Aroua M.K.; Raman A.A.; Sulaiman N.M.N. Methanol Recovery During Transesterification of Palm Oil in a TiO2/Al2O3 Membrane Reactor: Experimental Study and Neutral Network Modeling. Sep. Purif. Technol. 2010, 76, 58–63.

166. Wu H.; Zhang J.; Wei Q.; Zheng J.; Zhang J. Transesterification of Soybean Oil to Biodiesel Using Zeolite Supported CaO as Strong Base Catalysts. Fuel Process. Technol. 2013, 109, 13-18.

167. Babajide O.; Musyoka N.; Petrik L.; Ameer F. Novel Zeolite Na-X Synthesized from Fly Ash as a Heterogeneous Catalyst in Biodiesel Production. Catal. Today 2012, 190, 54-60.

168. Alves C.T.; Oliveira A.; Carneiro S.A.V; Silva A.G.; Andrade H.M.C.; Vieira de Melo S.A.B.; Torres E.A. Transesterification of Waste Frying Oil Using a Zinc Aluminate Catalyst. Fuel Process. Technol. 2013, 106, 102–107.

169. Borges M.E.; Díaz L. Recent Developments on Heterogeneous Catalysts for Biodiesel Production by Oil Esterification and Transesterification Reactions: A Review. Renew. Sust. Energ. Rev. 2012, 6, 2839–2849.

170. Di Serio M.; Tesser R.; Pengmei L.; Santacesaria E. Heterogeneous Catalysts for Biodiesel Production. Energ. Fuels 2008, 22, 207–17.

171. Wen Z., Yu X., Tu S.T., Yan J., Dahlquist E. Biodiesel Production from Waste Cooking Oil Catalyzed by TiO2–MgO Mixed Oxides. Bioresource Technol. 2010, 101, 9570–9576.

172. Guo F.; Peng Z.G.; Dai J.Y.; Xiu Z.L. Calcined Sodium Silicate as Solid Base Catalyst for Biodiesel Production. Fuel Process. Technol. 2010, 991, 322–328.

173. Singh A.K.; Fernando S.D.; Hernandez R. Base Catalyzed Fast Transesterification of Soybean Oil Using Ultrasonication. Energ. Fuel 2007, 21, 1161–1164.

174. Kalva A.; Sivasankar T.; Moholkar V.S. Physical Mechanism of Ultrasound-assisted Synthesis of Biodiesel. Ind. Eng. Chem. Res. 2008, 48, 534–544.

175. Nishimura CSMV; Maeda Y.R. Conversion of Vegetable Oil to Biodiesel Using Ultrasonic Irradiation. Chem. Lett. 2003, 32, 716–717.

176. Salamatinia B., Mootabadi H., Bhatia S., Abdullah A.Z. Optimization of Ultrasonic-Assisted Heterogeneous Biodiesel Production from Palm Oil: A Response Surface Methodology Approach. Fuel Process. Technol. 2010, 91, 441–448.

177. Mootabadi H.; Salamatinia B.; Bhatia S.; Abdullah A.Z. Ultrasonic-assisted Biodiesel Production Process from Palm Oil Using Alkaline Earth Metal Oxides as the Heterogeneous Catalysts. Fuel 2010, 89, 1818–1825.

178. Deng X.; Fang Z.; Liu Y.; Yu C. Production of Biodiesel from Jatropha Oil Catalyzed by Nanosized Solid Basic Catalyst. Energy 2011, 36, 777–784.

179. Deng X.; Fang Z.; Hu Y.; Zeng H.; Liao K.; Yu C.L. Preparation of Biodiesel on Nano Ca–Mg–Al Solid Base Catalyst under Ultrasonic Radiation in Microaqueous Media. Petrochemical Technology 2009, 38, 1071–1075.

Author Notes

CHAPTER 1

Acknowledgment
A.F.L. thanks the EPSRC (EP/G007594/3) for financial support and a Leadership Fellowship, and acknowledges the invaluable contributions of Prof Karen Wilson (European Bioenergy Research Institute, Aston University).

CHAPTER 2

Acknowledgement
The author would like to thank Ms Iva Simcic and InTech Europe for enabling her to publish this book chapter, while she is grateful to Mr Athanasios Dimitriadis who provided support, offered comments, proofreading and design. Finally she would like to express her appreciation for the financial support provided by the EU project BIOFUELS-2G which is co-financed by the European Program LIFE+.

CHAPTER 4

Acknowledgement
The results from the University of Wisconsin reported here were supported in part by the US Department of Energy Office of Basic Energy Sciences, by the DOE Great Lakes Bioenergy Research Center (www.greatlakesbioenergy.org), and by the Defense Advanced Research Project Agency.

CHAPTER 5

Acknowledgment
This material is based upon work performed through MOE Key Laboratory of Bio-based Material Science and Technology at the Northeast Forestry University, China and Department of Chemistry at Mississippi State University. This work was supported by t

CHAPTER 6

Acknowledgment
We deeply appreciate the financial support of Deutsche Bundesstiftung Umwelt (DBU) within the scope of this project.

CHAPTER 8

Acknowledgment
We are grateful to the National Natural Science Foundation of China (21072077), the Guangdong Natural Science Foundation (10151063201000051), and the Guangzhou Science & Technology Project (2010Y1-C511) for financial support.

Conflict of Interest
The authors declare no conflict of interest.

CHAPTER 9

Acknowledgments
EDC and PVDV thank Ghent University for financial support and Tom Planckaert from the department of Inorganic and Physical Chemistry (Ghent University) for nitrogen sorption and X-ray diffraction measurements. IDR and EMG acknowledge the support of the Région Wallonne (DG06, convention LIGNOFUEL No. 716721).

Conflict of Interest
The authors declare no conflict of interest.

Index